SpringerBriefs in Applied Sciences and Technology

SpringerBriefs present concise summaries of cutting-edge research and practical applications across a wide spectrum of fields. Featuring compact volumes of 50 to 125 pages, the series covers a range of content from professional to academic.

Typical publications can be:

- A timely report of state-of-the art methods
- An introduction to or a manual for the application of mathematical or computer techniques
- A bridge between new research results, as published in journal articles
- A snapshot of a hot or emerging topic
- An in-depth case study
- A presentation of core concepts that students must understand in order to make independent contributions

SpringerBriefs are characterized by fast, global electronic dissemination, standard publishing contracts, standardized manuscript preparation and formatting guidelines, and expedited production schedules.

On the one hand, **SpringerBriefs in Applied Sciences and Technology** are devoted to the publication of fundamentals and applications within the different classical engineering disciplines as well as in interdisciplinary fields that recently emerged between these areas. On the other hand, as the boundary separating fundamental research and applied technology is more and more dissolving, this series is particularly open to trans-disciplinary topics between fundamental science and engineering.

Indexed by EI-Compendex, SCOPUS and Springerlink.

More information about this series at http://www.springer.com/series/8884

Yasuhiro Suzuki • Katsushi Nakagawa
Takashi Sugiyama • Fuminori Akiba
Eric Maestri • Insil Choi • Shinya Tsuchiya

Computational Aesthetics

Springer

Yasuhiro Suzuki
Department of Complex Systems Science,
Graduate School of Informatics
Nagoya University
Nagoya, Japan

Takashi Sugiyama
Kyoto University
Kyoto, Japan

Eric Maestri
CICM (Centre de recherche en Informatique
et Création Musicale)
University Paris 8
Saint-Denis, France

Shinya Tsuchiya
ZA/ZA Aoyama
Tokyo, Japan

Katsushi Nakagawa
Institute of Urban Innovation
Yokohama National University
Yokohama, Japan

Fuminori Akiba
Graduate School of Information Science
Nagoya University
Nagoya, Japan

Insil Choi
Rohde Island School of Desing
Providence, Rhode Island, USA

ISSN 2191-530X ISSN 2191-5318 (electronic)
SpringerBriefs in Applied Sciences and Technology
ISBN 978-4-431-56842-1 ISBN 978-4-431-56844-5 (eBook)
https://doi.org/10.1007/978-4-431-56844-5

Library of Congress Control Number: 2018951586

This Springer imprint is published by the registered company Springer Japan KK part of Springer Nature.
The registered company address is: Shiroyama Trust Tower, 4-3-1 Toranomon, Minato-ku, Tokyo 105-6005, Japan

Preface

Computation and algorithms are broadly classified into two categories: machine computation and natural computation. Machine computation is the generic term for computation. In machine computation, the computing subjects are humans who use computers, such as PCs and abacuses, which are separate from the computing subject.

In natural computation, computing subjects and computers are integrated; hence, the computing subjects are computers. In machine computation, as computers exist outside the computing subject, computation is completed by the computers; hence, the subjects are simply waiting for the computers to provide their result. Machine computation is essentially mathematics, and the computers that exist outside of the computing subjects have been investigated in detail.

By contrast, in natural computation, as the computing subjects themselves are also computers, the subjects must decide when to start and halt computation. Consequently, computations through natural computation sometimes do not halt. For example, the "natural computation" of the French artist Alberto Giacometti while expressing his impression of Japanese philosopher Isaku Yanaihara did not halt easily and took several months; even if "materials for computation" such as canvas or clay could not be used (because of too much paint or sharpening sculpture), computations continued without a pause. If this scenario were to occur with machine computation, the computers would break and be physically forced to stop.

Natural computation is, in a broad sense, essentially aesthetics, where it is not limited to art and includes preferable, comfortable, stable, … (a set of words with nature and appreciation of beauty here) states. This very long list of states of beauty encompasses the essence of natural computations and provides various algorithms that can lead the computing subjects to halt computation.

When I started thinking about natural computation, I did not expect that the matter of aesthetics could encapsulate the essence of natural computation. This book gathers discussions and practices on computational aesthetics from philosophy, aesthetics, art, and computational science. The specialty fields of the authors range from philosophy, aesthetics, and computational science to hair design, scientific illustration, sound art, and composition.

Yasuhiro Suzuki argues computational aesthetics from computer science and Kansei information (tactile interactions and communications) point of view, Fuminori Akiba discusses on the significance of natural computing for considering computational aesthetics of nature, Takashi Sugiyama argues Herder on "Sentio, Ergo Sum" from his remarks on the color Harpsichord, Katsushi Nakagawa provides a historical perspective on sound art and its relationship to tactile sense, Eric Maestri discusses (composer) on a sector-gestural-morphological analysis of a musical-tactile score, and Shinya Tsuchiya (hair designer) analyzes hair design as algorithm and shows computational aesthetics of hair design.

Every chapter is thought-provoking, and the reader will realize that the field of computational aesthetics is spreading in various directions and will increase its popularity.

Computational aesthetics has been a "fragile" concept but has been strongly supported by many braves; we would like to express thankfulness for all. When I was not confident of this fragile concept, Prof. Yukio-Pegio-Gunji (Waseda University) asked me to give intensive lecture and gave chance to open up; this lecture was a turning point to start developing this crazy idea (a reader who want to know "how crazy it is" please proceed reading). Professor Fuminori Akiba (Nagoya University) has also encouraged through thoughtful discussions. Long discussions with Insil Choi have made us notice various aspects that we have not realized. Professor Kazuhisa Todayama (Nagoya University) has been stimulating from the viewpoint of philosophy of computation. The invited lecture of Prof. Ricard Shusterman (Florida Atlantic University, Prof. F. Akiba invited him) taught us the significance of this concept, and we would like to express our gratitude to Prof. Shigetoshi Osano (Honorary professor of Tokyo University). Constructive suggestions and warm encouragement from relevant aestheticians have strongly encouraged our projects.

This book involved a considerable amount of time and effort, and I am deeply grateful to every author who patiently cooperated and for the editor's kind support. This publication has been supported by the grant in aid for scientific research No. 15K00268, 15K02105, 24520106, 23300317, and 16H03093; interdisciplinary discussions supported by Knowledge Key Knot Keepers – Designated Academic Network, KKKK-DAN.

Nagoya, Aichi, Japan Yasuhiro Suzuki, on behalf of authors

Contents

Chapter 1
Computational Aesthetics

Yasuhiro Suzuki

Abstract Aesthetics is the most fundal concept in Natural Computing; Aesthetics is knowledge itself and it can only halt computing in Nature. We redefine Computations and show that Aesthetics is irreducible concept in Natural Computing. We consider the Computational Aesthetics in visual by using generative arts. In order to investigate Computational Aesthetics in tactile sense, we propose Tactile Score and propose a design principle of Aesthetics of tactile sense based on investigation of massaging. The principle is verified by through examining biological responses of Stress marker and Brain activities when rationally designed tactile sense is given.

Keywords Tactile Score · Natural Computing · Generative Algorithmic Art · Genetic Algorithm · Massage · Effectiveness of massaging

1.1 What Is Computation?

Mathematically, the name of algorithm, comes from Arabic mathematician "*Al-Khuwarizmi*" (ninth century); he proposed a method of solving an equation as sequences of instructions, it was an origin of Algebra and algorithm (Greek mathematician, Euclid proposed an algorithm to obtain the greatest common divisor of two numbers, in BC 300), in general; in Computer science, the definition of algorithm was given by Turing Machine (1936), which was proposed by Alan Turing (1912–1954). TM was the beginning (big bang) of computer science and it is also composed of sequences of instructions, described with mathematically strict format (Menninger 1969).

After TM, the word of algorithm means that an algorithm for computer, however it has proposed since 9th Century (or BC 300), algorithm is not only for artificial

Y. Suzuki (✉)
Department of Complex Systems Science, Graduate School of Informatics,
Nagoya University, Nagoya, Japan
e-mail: ysuzuki@nagoya-u.jp

Y. Suzuki et al., *Computational Aesthetics*, SpringerBriefs in Applied Sciences and Technology, https://doi.org/10.1007/978-4-431-56844-5_1

languages for computers (e.g. programming language like Java, Python, etc.) but also a broad method to perceive the nature and ourselves.

Algorithm is a method of describing the sequences of processing something; so, brushing teeth can be described by algorithm as;

1. put tooth paste on a brush,
2. brush teeth,
3. rinse off tooth paste;

programming is (re-)arrangements of the sequences; for example, we will obtain another algorithm by programming as follows, where we exchange 1. to 3.;

3. rinse off teeth,
1. put tooth paste on a brush,
2. brush teeth;

so, if this algorithm is executed, full of bubbles of tooth paste will be filled inside the mouth; where every process is exactly the same as the former algorithm and only the order of execution of each process is different. Hence, a way of drawing a picture, a cooking recipe, tell a person how to go to the destination, the methods of teaching, etc. are algorithms.

1.2 Algorithm Brings Miracle

Almost all the processes in Nature or daily life are not dis-ordered but well-ordered, and the order has been called Algorithm; if there is a tomato, a bottle of olive oil, pasta and some salt; if there are not algorithm, nothing will happen, while if an Italian chef gives a good recipe; they are transformed into *spaghetti al Pomodoro* and we can reproduce the *spaghetti al Pomodoro* as much as we want; it never happens just see these materials for infinite time; algorithm has brought a plenty of miracles from the origin of things; if an algorithm is wrong, quality of products would be not good or nearly impossible to produce the aimed product.

1.3 Algorithm + Executer = Computing

An algorithm is just a sequence of instructions and cannot do anything by itself, in order to compute something, it requires "someone" as a programmer, user, etc. and "something" to execute the algorithm as computers, machines, chemical systems, etc.; we will call the "someone" as main constituent of computation, MCC and "something" as executer (Fig. 1.1).

Fig. 1.1 External Main Constituent of Computation, MCC and an Executer, which has algorithm to execute

Fig. 1.2 Internal Main Constituent of Computation: where the Executer is MCC, which has algorithm to execute

There are two types relationships between MCC and algorithms as follows;

Internal MCC: in this case MCC are executers so MCC have to decide when computations halt themselves; cooking, drawing correspond to this case (Fig. 1.2),

External MCC: in this case MCC exist in the outside of executers, so once computations start, MCC wait for the computation halt (Fig. 1.1).

1.4 Type of Algorithms

I distinguish two sides of algorithm below as follows;

Type I: a method to perceive the world; broadly applicable and which is not limited only by using computers; not only Algebra but also cooking, teaching, brushing teeth, etc. are based on this type of algorithm;

Type II: algorithms for computers, it is being noticed that Type I and Type II are not completely divided, Type I will give birth of Type II, *vice versa*; we perceive something and in some case, try to describe by using Type I and transform into Type II then observe computing in the executer such as computer simulations; and the simulations give hints to explore novel Type I algorithms.

We have been investigating Type I and transformed into Type II; hence consequently techniques described by Type I algorithms, for example, to create art works must be transformed into Type II; it has been difficult to treat Type I algorithm for non-specialists and artists who possess such techniques have formed a "guild" and sustain their "position" in a society; however such guilds will be disappeared by Type II algorithm and computers,; for sure, photographers have been losing their jobs and young digital-agers will not know anything about a camera film.

As Ray Kurzweil (Kurzweil 2005), who predicts technological singularity will come in the near future, has pointed out that computer science, especially Artificial Intelligence develops dramatically; Type I will be transformed into Type II much more rapidly than ever, so everyone will be able to obtain various techniques to create art works very easily that have been permitted to treat only well talented and trained artists.

The number of Type I algorithms will decrease; jobs that have stronger tendencies of relying on Type II algorithm such as judicial affairs or medical diagnosis have already been taken over by artificial intelligence, gradually (but at the moment) and such Type II tending professions would be deprived by computer systems in the near future.

However, Type I algorithms never disappear; because "our" creativity creates Type I algorithms infinitely; where "our" means that the whole algorithm-useable matters in the nature broadly, are not human beings only.

1.5 Aesthetics in Computing

In the case when you are a MCC and cocking a dish according to the recipe of *spaghetti al Pomodoro*, you will start an execution in the algorithm; for example, "frying minced onion until its color turns to brown" then you have to decide when stop frying the onion; the recipe would not instruct the condition of stop frying by the required color spectrum distribution; in many recipes that instruct "adjust taste with salt and pepper" so we have to decide when stop adding salt and pepper; the recipe does not tell you the required concentration of NaCl. In Type II algorithm, we must define the condition of halt computation, otherwise computations will never halt.

On the other hand, in Type I algorithm a MCC can halt computing in anytime, in other words, computing never stops if the MCC continues computation. What condition will let a MCC decide halting computation? Let us consider the case when you adjust taste adding some salt to soup, how to decide to stop adding salt? Surely most of you, taste a small amount of soup then decide the quantity of salt to add;

how can you know, should not add more salt? Probably it is because the taste of soup is good, this goodness let MCC (you) halt adding salt; as this example, in Type I algorithm, goodness, comfortableness, beautifulness, etc. such matters of aesthetics decide the condition of halting computing.

Hence in some cases, computing does not halt; Alberto Giacometti (1901–1966), sculptor tried to create an art work of Japanese philosopher, Isaku Yanaihara (1918–1989), it took almost 1 year; he tried to various methods of expressing, drawing, oil painting, sculpture, etc. and in some time, created artwork was disappeared by too much colors, lines etc. or too much removing soil from the sculpture; it shows the case when the executed algorithm does not halt because of the computing state does not reach to the halt state defined by his aesthetics (Yanaihara 2010).

This Giacometti's case seems an extraordinary example, but we did almost same things in daily life; when you are preparing a soup for a formal party and if it tastes incredibly bad, you will throw away it and start preparing another soup again or look for canned soup; in case, you are invited to a special dinner; you will make up very carefully and even if you take several ten minutes to make up, if your aesthetics cannot accept it, you will throw away it and start making up again.

Therefore, aesthetics is principally irreducible for algorithm of Type I and one of the most important characteristics, the condition of halt computing relies on the aesthetics of a MCC, even if we do not consider about art or culture.

Aesthetics is not a terminology, which is allowed to use only in Art or Philosophy, Aesthetics is the most important instinct to survive for MCCs in wild life including humankind; it has well known that flies have the sense of taste and of course they like "tasty food sources". Their tasty foods are not tasty for humankind, whether something is tasty or not is closely relating to the safety (Choi n.d.); it has not been confirmed there exists a school for children of fly or a mother of fly teaches her children how to distinct harmful food sources; where Aesthetics of taste teaches, which is safety food.

Aesthetics is not only for wild life, if a protein molecular does not have a kind of tactile sense, proteins will not be able to produce functions; they have a kind of tactile sense and when a protein touches to another protein, if it is tactually "good", they will connect with each other, otherwise does not connect; protein's Aesthetics of tactile sense produces preferences of binding proteins and the preferences produce functions of proteins.

1.6 Aesthetics in Arts

Needless to emphasize, Aesthetics is matter of beauty and closely related to (fine) arts; so, what is Arts? *Art is explosion*, which is a definition of Art by Taro Okamoto, Japanese artist (1912–1996); he did claim that art eliminates boundaries of ourselves and societies, it does not make any sound and while people cannot sense anything, arts completely eliminates every boundary and promote re-built our human and society (Choi n.d.).

In order to make explosion, he proposed three basics in art;

1. Art must not be comfortable,
2. Art must not be fine,
3. Art must not be created by matured good skills,

he strongly claimed that true beauty is never be just fine, and matured artistic skills prevent an art work from true beauty, such an art work is just fine, just good looking, just good for decorating a fancy room, but never makes any explosion.

Rieko Suzuki (CEO, Face Therapy Co. Ltd.) who is a Japanese haptic artist has given almost same definition for arts; she claims that art is knowledge, which directly evokes and immediately inserts our senses deeply while has not known and avoids to be expressed by language or logic; a real art work never requires explorations by language or logic; a real art is violently forced us to know it through our senses and after this "initiation evoked by explosion" then our intelligence which is running by language and logic realizes as it seems ... there may exist something important (Suzuki n.d.-a).

Concerning with modern arts in these days, prof. Fuminori Akiba (Nagoya University), aestheticians share the definition of arts and critiques "*so-called most of modern arts are not real arts" and also most of all the artists miss-understood about what they are doing, some artists criticize scientists and they claim that arts have proceeded toward science and we artists have been leading science"* he points out "*certainly such arts existed, but they were in* 17^c *or* 18^c "; and professor Akiba claims "*in this science era, good scientists are real artists and so-called artists are just followers of science; they must realize it and must wake up"* (Akiba n.d.).

1.7 Computational Aesthetics

Aesthetics boasts long research history and it is interesting that Since 18 Century after Immanuel Kant, considering Aesthetics philosophically has been "banned" by his critique and aestheticians became historians (Akiba n.d.).

On the other hand, when we observe nature algorithmically (from Natural Computing view), Aesthetics is the fount of Natural Computing; it is the origin of basic interactions such as every force and preferences in nature and it decides halt of computing; when the process reached "Aesthetic state(s)", the processes of natural computing halt; Aesthetic state(s) are denoted as *beautiful, comfortable, tasty, sate,* etc.; as we addressed in Giacometti's case or cooking dishes, Aesthetic state depends on the type of computing.

Hence, Aesthetics is not only a matter of Philosophy but all about Science (not limited in Natural Science but including Humanities, Social Science, Fine Arts and liberal Arts, etc.). Computational Aesthetics considers "what is Aesthetics" in algorithmically (Suzuki n.d.-b).

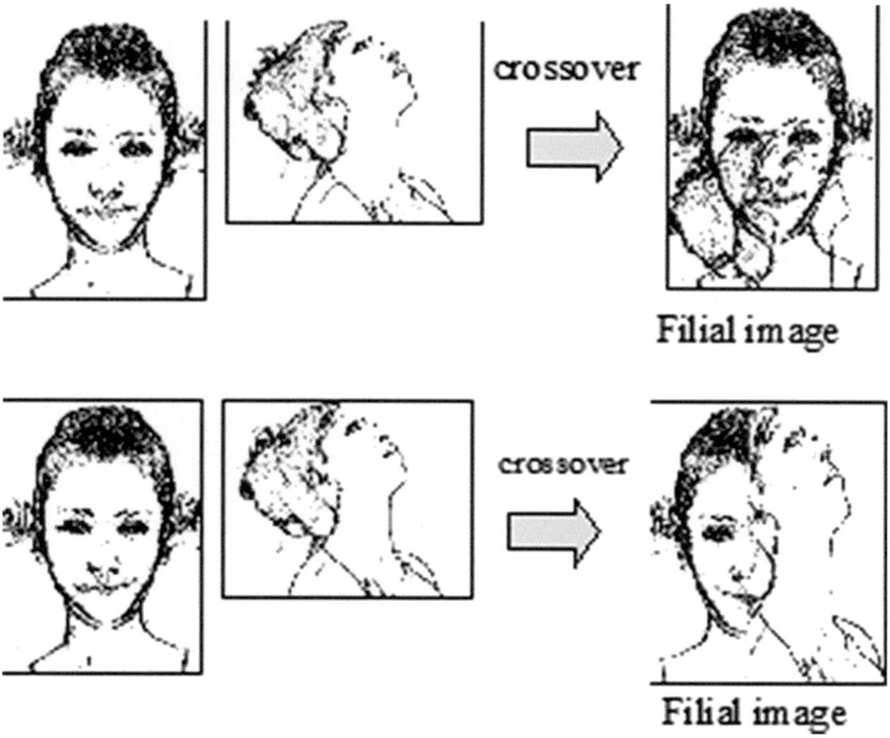

Fig. 1.3 (above) Example of creating a filial image by crossing over two pictures, (below) by crossing over of vertically partitioning pictures

1.8 Arts by Algorithm

In order to investigate people's Aesthetics for visual arts, we did experiments by using gradually changing paintings and asked people which is interesting or preferable (Suzuki and Suzuki 2014a); we have created pictures by using Genetic Algorithm (GA), which is a biomimetic algorithm based on Darwinian evolution. At the beginning, the final goal of picture is set, which can be objects as a face, apple, triangles, etc. or basic composition of visual art as triangular, perspective, etc. Then evolution has started from simple line drawings of concrete objects, such as a person's face, scissors, apple, etc.; two pictures are selected as a "mother" and "father" picture and their "child" picture is obtained by crossing over their pictures (Fig. 1.3).

1.8.1 Mutation and Selection

Mutation: In order to implement Darwinian evolution, we implement mutation and selection; mutation is implemented by randomly reversing the color of pixel between black and white; the mutation threshold is set and we assign a random number to

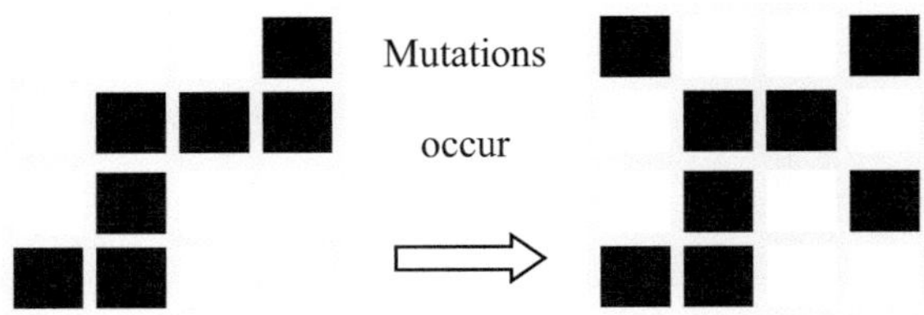

Fig. 1.4 Mutation; in this example, the leftmost at the top pixel is mutated and reversed its color from white to black and in the middle of rightmost at the bottom, reversed from white to black

each pixel and the pixel of which random number is below the threshold is reversed its color between black and white (Fig. 1.4).

Selection: The next generations of filial pictures are selected by the fitness of a filial picture, which is given how the filial image is closer to the final goal of pictures; it is obtained by counting the number of overlap pixels; we can select only pictures having higher fitness (elites) for the next population; if we select only elites for the next population, the diversity of pictures will decrease and they would be only similar pictures, while if we ignore fitness and just select pictures randomly, although the diversity of pictures would be maintained, the evolution process would be difficult to be converged.

1.8.2 Arts by Algorithm

The following artworks are obtained by random selection and elite preservation. In both trials, we used four pieces of initial images (Fig. 1.5a) and one final goal picture (Fig. 1.5b).

In case we take the random selection, even if an individual filial picture with the high fitness, there is no guarantee such an elite filial picture will be handed over to next generation hence the initial pictures can't get closer to the target image but the diversity of generating pictures to be kept (Fig. 1.6). It is shown in the time development of fitness value (Fig. 1.7); we examined the development of fitness value during evolution; compared the average fitness of whole population with the elite population; they were similar and had no significant differences (Fig. 1.7).

On the other hand, in the case when elite pictures are preserved, the produced pictures were getting closer to the target image (Fig. 1.5a). Comparing to the images produced by random selection, there was less image noise, and static images tend to be output and generated images are resembling the final goal picture and evolution processes to be converged.

The number of individuals in each generation, mutation rate, image size, and the score are equivalent in the preceding section when the initial image matches or mismatches with the target image. The number of preserved elite individuals in each generation is 5 (Fig. 1.8).

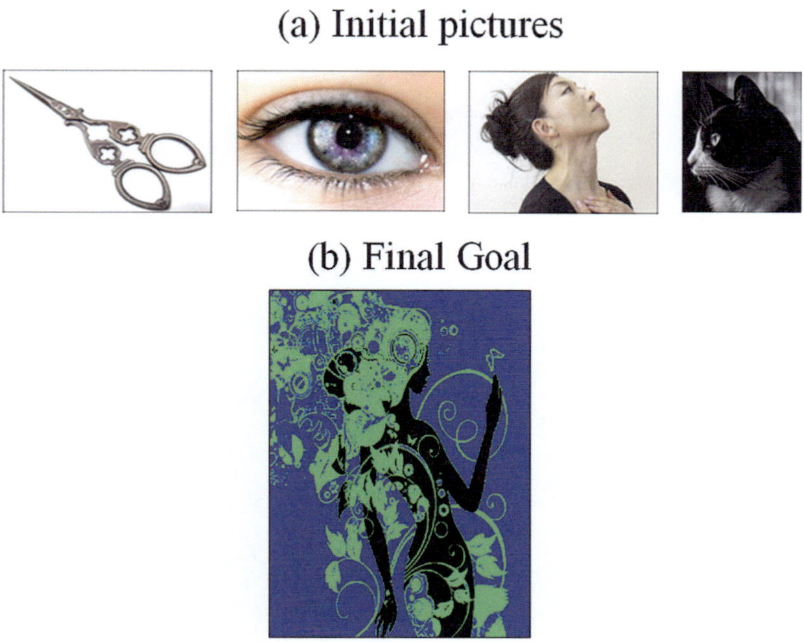

Fig. 1.5 (**a**) Initial pictures for trials; these initial pictures are converted into line drawings; (**b**) picture of final goal; we converted the background to white and transformed into a line drawing

In the elite preserved selection, if an individual filial picture with the high fitness, it will hand over to the next generation hence the initial pictures can get closer to the target image, while the diversity of generating pictures is decreased (Fig. 1.8). It is shown in the time development of fitness value (Fig. 1.9); we examined the development of fitness value during evolution; compared the average fitness of the whole population with the elite population; it shows the fitness values of the elite population is significantly higher than average.

1.9 Experimentally Investigation of Aesthetics of Visual Art

In order to investigate people's aesthetic of visual art experimentally, we generated artworks by using this system and examined impressions by using methods of basic Psychology.

We used the same set of initial images, with elite preservation and prepared two different goal images, a clover leaf and standing lady (Fig. 1.5) and the triangular composition (Fig. 1.10); we used generated artwork in early stage, in the middle of convergence and converged. We obtain three different strains of evolution of artworks; all the strains start from the image of objects and they evolve to abstract, triangle and nonobjective images.

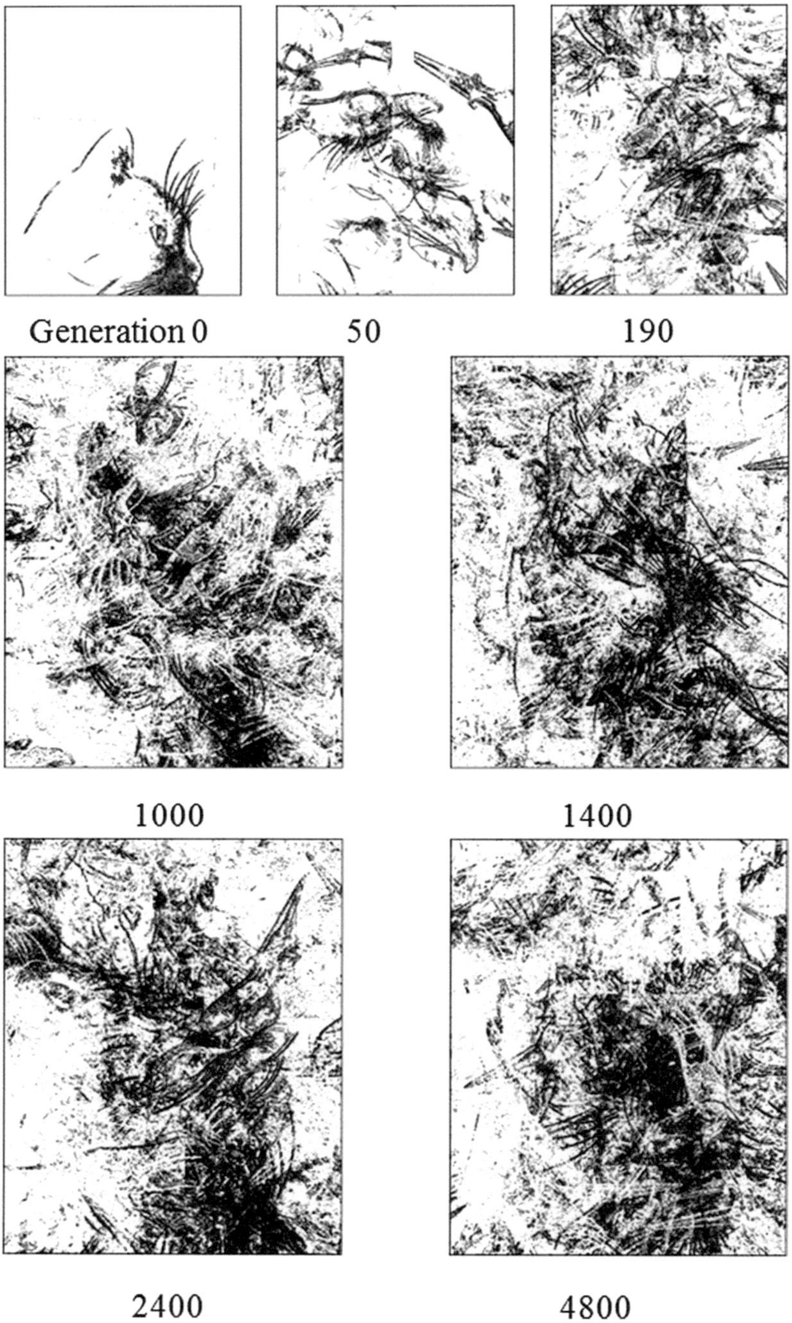

Fig. 1.6 Output results without elite preservation, where the number of generation denotes development time in evolution processes from generation 0 (initial) to 4800 generation; one generation means that it generates population of filial pictures by crossing pictures and selecting filial pictures randomly for the next population: The number of individuals in each generation n = 80, the mutation rate is 0.000005, picture size is 600 × 800 pixels

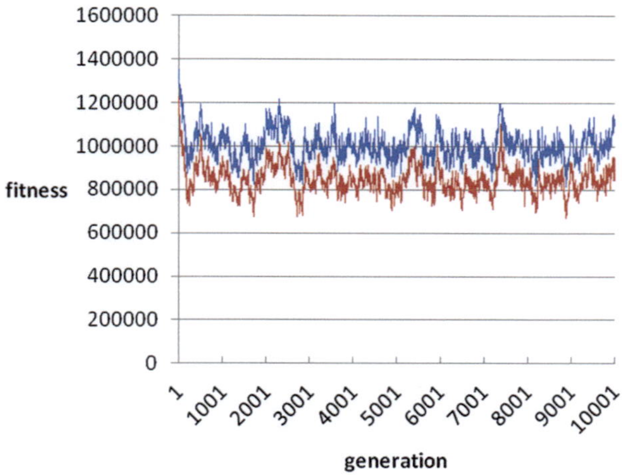

Fig. 1.7 Graph of the fitness of each generation; red line illustrates the average fitness of the whole population, blue, the elite population

The questionnaire was conducted at the booth of the *creator's market* vol. 2914 in *Port Messe Nagoya* from 7th to 8th December 2013. The *creator's market* is a big event that attracts more than approximately 3000 people including professionals and amateurs for selling all sorts of original works such as fashions, interiors, crafts, visual designs, etc. We joined this event and got answers from 19 male and female participants between teenagers.

The results of the questionnaire are shown as points of Semantic Differential, SD method. We prepared three sets of generated images in early, middle, and later stage. Each image was printed on a piece of paper, and they were arranged in random order.

In the questionnaire results, high values are observed in early and middle generation, especially dynamic motion in middle generation shows extremely high values (Fig. 1.11). The three items "pleasure", "fun", and "originality" in early generation exceed those in middle generation, but the item "beauty" is at the same level. All the items in later generation are below those in early and middle generation, and especially dynamic motion is low.

1.9.1 Factor Analysis

We describe the results of the factor analysis of the questionnaire. We got the cumulative contribution ratio and eigenvalue plot from the factor analysis without deciding the number of factors. The number of factor fields is numbered in descending order of the contribution ratio for descriptive purpose. If the number of factors is determined using the Gutmann standard and Scree standard, the number of factors becomes excessive, so it is difficult to grasp the structures. Therefore, we employ top two factors that their eigenvalues are more than 3 and have extremely high contribution ratios in order to facilitate the interpretation. We conducted the factor

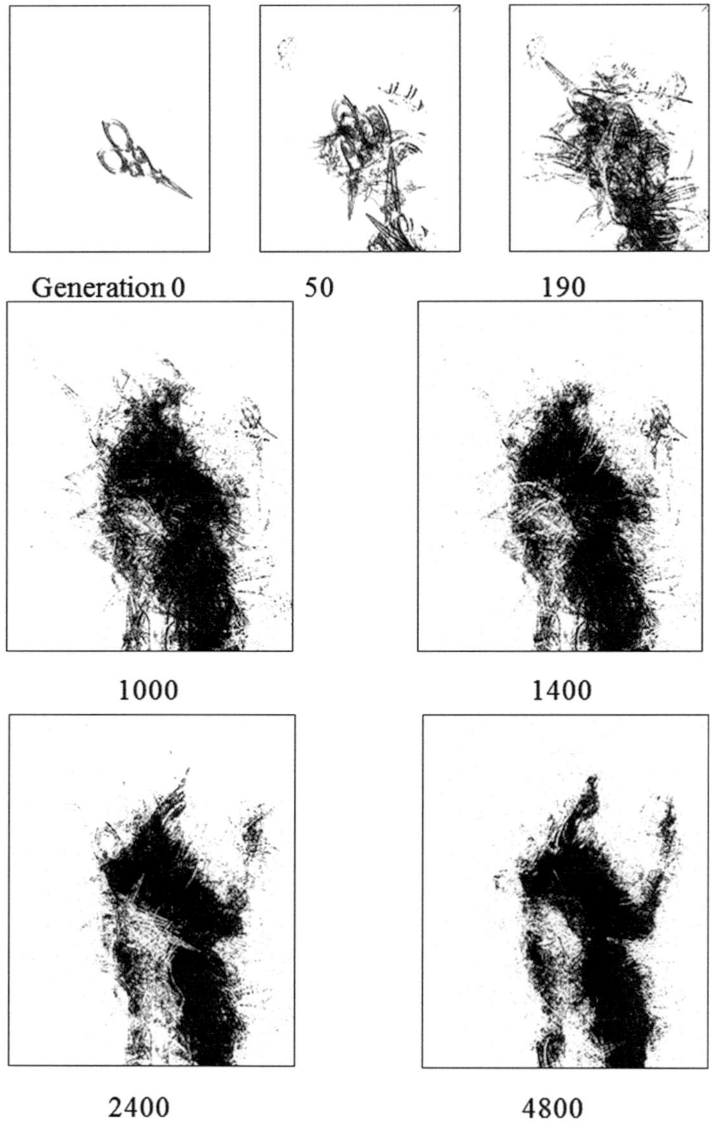

Fig. 1.8 Output results of elite preservation; where the number of generation denotes development time in evolution processes from generation 0 (initial) to 4800 generation; one generation means that generate population of filial pictures by crossing pictures and select elite filial pictures for the next population: The number of individuals in each generation n = 80, the mutation rate is 0.000005, picture size is 600 × 800 pixels

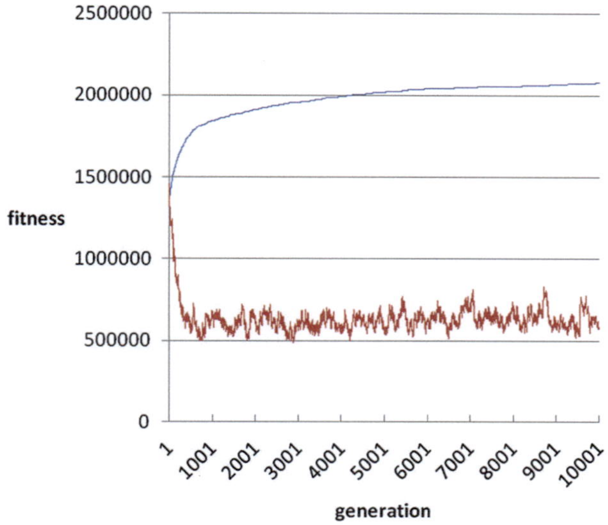

Fig. 1.9 Graph of the fitness of each generation; red line illustrates the average fitness of the whole population, blue line, the elite population

analysis of these two factors using the maximum-likelihood method and pro matrix method. We found that the first factor had high factor loadings related to aesthetic evaluation of the output picture such as pleasure and beauty in early and middle generation, and then we named this factor as "aesthetic evaluation" factor.

The second factor had extremely high factor loadings related to fun and originality in early generation, the second highest factor loading was dynamics in early, middle, and later generation, and fun in later generation except for pleasure and beauty. These are related to originality and movement of pictures, so we named this factor as "originality" factor (Table 1.1).

1.9.2 Multiple Range Test

We conducted multiple range test on each item in the results of the questionnaire and examined the significant difference (Table 1.2). The results show 1% significant difference between early and later generations on pleasure items, between early and middle, and between middle and later generations on dynamic items, and between early and later generations on originality items. Similarly, there were 5% significant differences between early and later generations on beauty items, between middle and later generations on favorability items, between early and middle generations on fun items, and between middle and later generations on originality items.

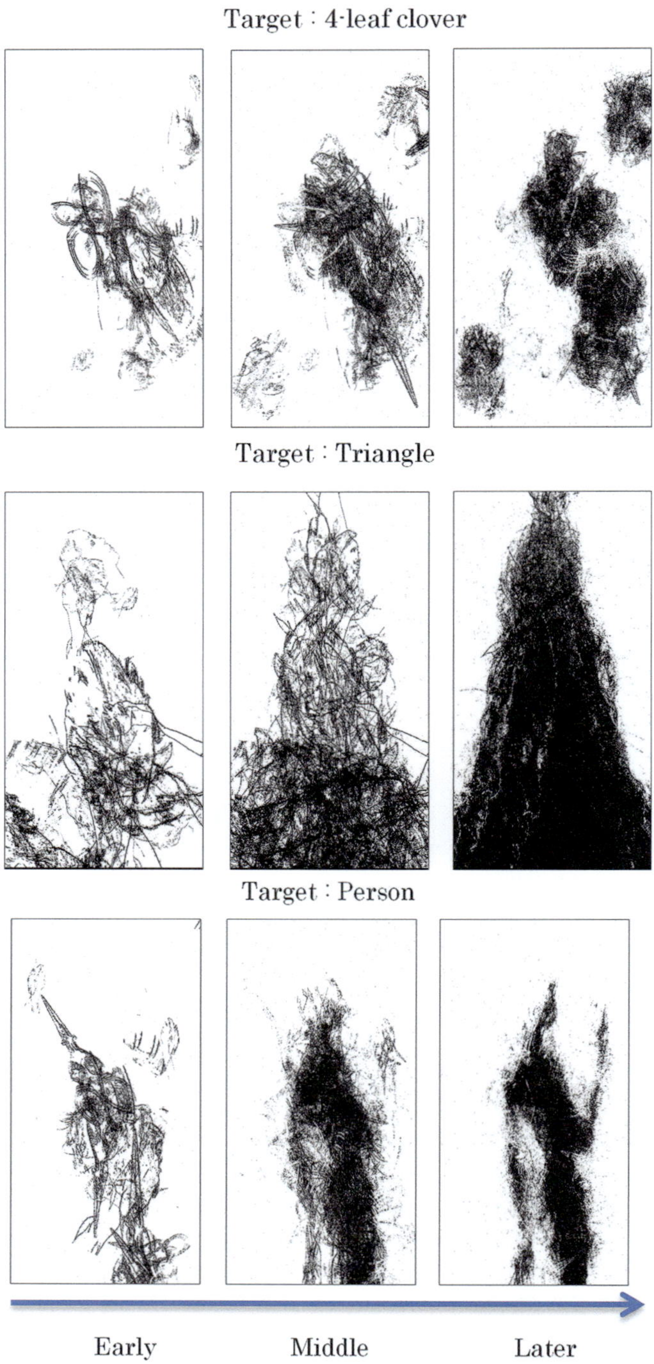

Fig. 1.10 Images used for the questionnaire

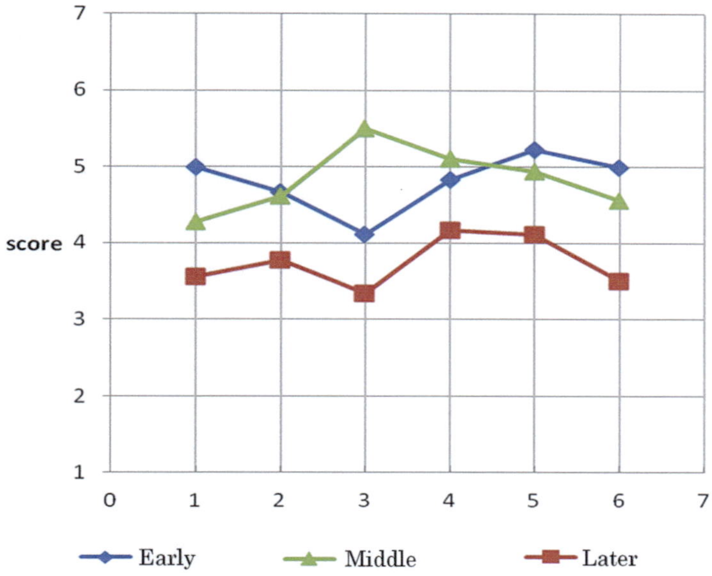

Fig. 1.11 Average values of the questionnaire results

Table 1.1 Results of factor analysis

	Factor 1	Factor 2
(Early)		
Comfortable	0.5965	0.3899
Beauty	0.5411	0.4548
Dynamics	0.1918	0.4923
Pleasure	0.382	0.5761
Fun	0.1238	0.7694
Originality	0.0904	0.6835
(Middle)		
Comfortable	0.7263	-0.0682
Beauty	0.7626	-0.1485
Dynamics	-0.163	0.4064
Pleasure	0.6504	-0.2494
Fun	0.8858	0.0162
Originality	0.7779	0.0534
(Later)		
Comfortable	-0.2132	0.1646
Beauty	-0.5756	0.5579
Dynamics	-0.2353	-0.539
Pleasure	-0.4763	0.3311
Fun	-0.4645	0.4182
Originality	-0.1099	0.3138

Table 1.2 Results of the multiple range test

(e); Early, (m); Middle, (l); Later		P value	
(e) Comfortable	(m) Comfortable	0.2733	
(e) Comfortable	(l) Comfortable	0.0014	**
(m) Comfortable	(l) Comfortable	0.0885	
(e) Beauty	(m) Beauty	0.9697	
(e) Beauty	(l) Beauty	0.0316	*
(m) Beauty	(l) Beauty	0.0556	
(e) Dynamics	(m) Dynamics	0.0056	**
(e) Dynamics	(l) Dynamics	0.3673	
(m) Dynamics	(l) Dynamics	0.0001	**
(e) Pleasure	(m) Pleasure	0.8577	
(e) Pleasure	(l) Pleasure	0.1284	
(m) Pleasure	(l) Pleasure	0.0394	*
(e) Fun	(m) Fun	0.8228	
(e) Fun	(l) Fun	0.0085	**
(m) Fun	(l) Fun	0.0401	*
(e) Originality	(m) Originality	0.6044	
(e) Originality	(l) Originality	0.0025	**
(m) Originality	(l) Originality	0.0347	*

1.10 Result of Experimental Aesthetics

The results of the questionnaire with elite preservation show that they came out of the opposite of the model fitness. From the table of actual factor analysis, it is believed that the fitness evenly becomes higher at every generation, on the other hand both originality and aesthetic evaluation are high in early generation, and then the originality comes down but the aesthetic evaluation goes up at the generations roll by. And the aesthetic evaluation decreases in accordance with the late generation.

These results show that interesting images are output in early generation whose fitness is low or in middle generation whose fitness starts to increase. Therefore, the model that simply evolves and aims at the target image is inappropriate. This indicates that the model that sustains the fitness in early and middle generations can output better images. The models that stochastically conduct elite preservation or change points and constrain its evolution at the time of coincidence of images when the fitness becomes excess during evolution are good examples (Fig. 1.12).

The results show that the case without elite preservation keeps the fitness low, but a wider variety of output images are created than those in the case with elite preservation. When Fig. 1.13 is taken as an example, it evolves without elite preservation using a four-leaves clover picture as a target image, the output image becomes like a witch who looks to the distance, so there is a possibility to output a heuristic image from the target image beyond all imagination.

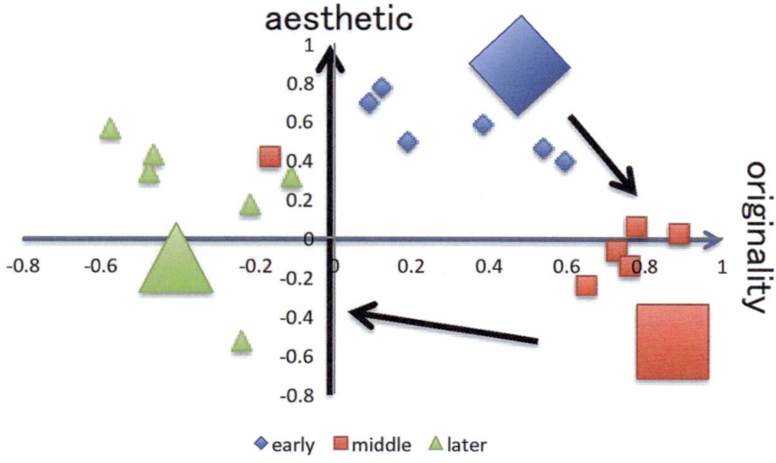

Fig. 1.12 Schematic diagram of the changes of evaluation by generations

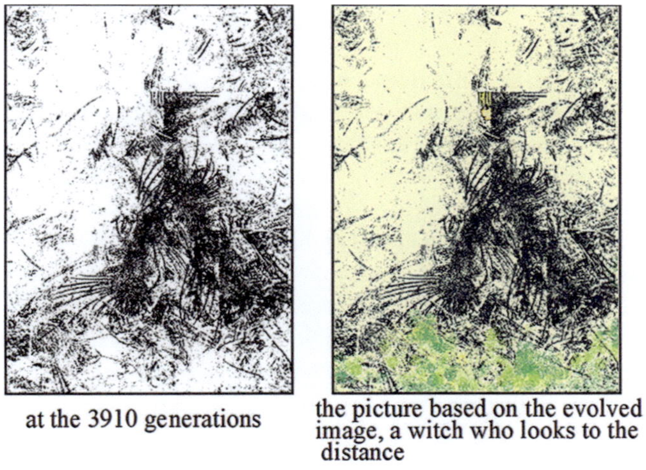

Fig. 1.13 Pictures based on an evolved image

These results suggest that when the fitness in this model is employed, the area where we can expect an image having high aesthetic evaluation, originality, and heuristic output (Fig. 1.14).

These are rough considerations, but we could show that extremely simple models in this study can output various pictures. These results suggest that it is worthwhile creating a model that passes through the above area and carrying out further research.

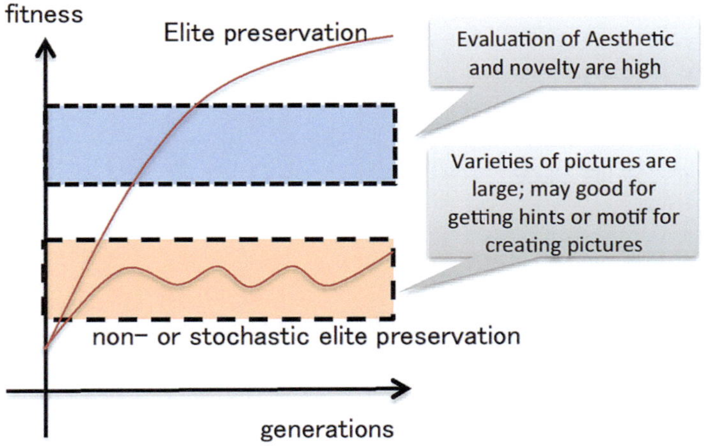

Fig. 1.14 Schematic diagram of each area

1.11 Aesthetics of Tactile Sense

We share the same tactile sense from a person or architecture that can or cannot get along. Like the atmosphere produced by architecture, we cannot see and hear the sense of touch but we can feel its existence strongly.

What is Aesthetics of tactile sense? One of an art of tactile sense, which we have developed for more than 2,000 years, is massage; a massage is composed of various tactile senses that are mainly generated by hand movements, in which the sequence of tactile senses is essential and we are able to compose the tactile sense by designing the sequence of hand movements. Hence, Aesthetics of massage is composed by the sequence of tactile senses, so we are able to investigate Aesthetic in massage by analyzing the principle of composing massages.

1.12 Tactile Score

In order to investigate massages, we have developed a notation for describing massages, the Tactile Score® (Suzuki and Suzuki 2014b); we borrow a notation of music score. We have found that a massage is composed of three main basic elements; they are pressure, contact area, and rhythm; hence by describing these three elements, we are able to describe tactile sense of massage precisely.

The pressure of massage varies from person to person, so we describe the degree of pressure relatively; in order to that, we set the basic pressure, which is obtained by "calibration" of pressure; the calibration has been performed by imaging the pressure when we hold something very precious things such as a new baby; we hold the baby with neither very strong pressure nor weak pressure.

And the contact area of hands is important to express tactile sense, for example small contact areas such as fingertips give strong pressure like picking at a face with a stick, or palms of hands can express tactile sense as if a person wraps up a face.

In the Tactile Score, the third line illustrates the basic pressure, which is notated by Alto clef and below the third line (low-pitched sound) describes stronger pressure than the basic pressure, while above it is weaker than the basic pressure; Here the beat is quadruple time, but triple and double time are also acceptable.

Next, we number the areas of a hand to describe the size of the dimension (Fig. 1.15, upper), in addition, we encode the spatial position and the movement of the stroke like a curve, line, dot, and each size of them like small, medium, and large as tactile steps like sol-fa of a musical score on a face (Fig. 1.15, lower).

When a human hand creates tactile sense (Menninger 1969), the contact area corresponds to the touching dimension of the hand. We describe the rhythm in the same manner as a musical score.

In the tactile score of Fig. 1.16, at the first count in A_5 part, circles are drawn on both sides of the cheeks using the center of the palm with weaker pressure than the basic pressure, at the second count, the hands are moved to the tails of the eyes and small circles are drawn using the center of the palm while keeping the same pressure as the first count, at the third and fourth counts, the hands are moved to both sides of the cheeks and circles are drawn using the fingertips with a stronger pressure than the basic pressure.

1.12.1 Language of Tactile Sense

Tactile perception conveys different messages from speech language. When one is patted on the shoulder once, he/she might think of accidental collision, yet when patted twice, it has meaning and he/she interprets it as being called. Also, mothers gently tap their babies at steady rhythm in caressing; the steady rhythm evokes the sense of security in the babies.

In other words, counts and rhythm are important in tactile perception; a single circular stroke could not be distinguished from a mere rubbing, while more than double strokes would be recognized as massage. So, this "(more than) double strokes" is an "alphabet "of a language of tactile sense and a set composed of these alphabets is a "word"; a massage is composed of these words as if a sentence. A poem is composed of sentences and these sentences generate "rhythm" likewise a poem, sentences composed of tactile sense words also generate rhythm.

As mothers' gently tap, steady rhythm adds meaning and sense of security to massage, so such steady rhythm would be considered as measures in music. Empirically, we have found that many subjects like massages composed of the Tactile Scores with quadruple measures, so when we compose a Tactile Score, basic elements in 4 counts are used as one unit of massage.

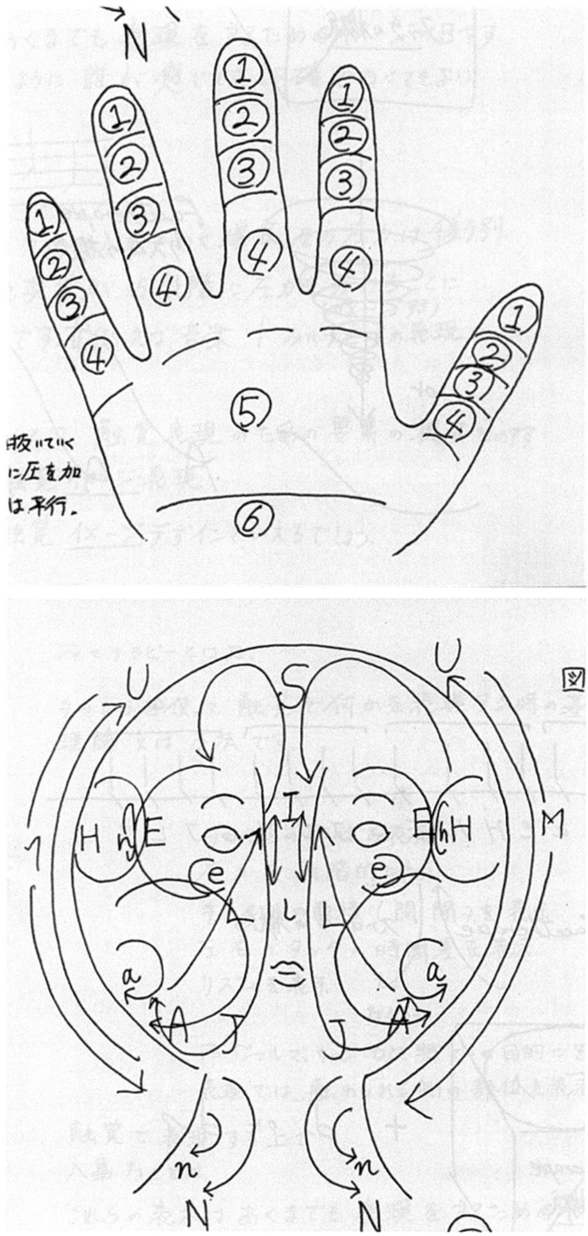

Fig. 1.15 Upper: Usage of part of the hand. Lower: Strokes of massage on a face; these strokes are obtained from massage experiences in aesthetic salons; strokes that pass uncomfortable areas are excluded

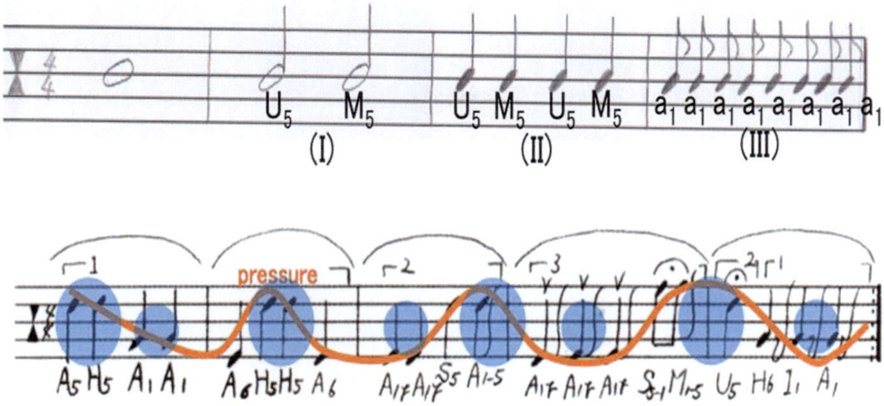

Fig. 1.16 Upper: Example of a tactile score, with special marking above the staff notation; Lower: 1 denotes both hands moving at the same, 2 indicates a small gap between hands, and 3 indicates a large gap between hands, the Sulla like marks illustrate a unit component of massage, the integral-like marks illustrate releasing pressure, and the breath-like mark corresponds to a short pause in massage, much like a breath in playing music

By giving a rhythm on a tactile sense, we can create "impressions"; a rhythm of touching gives a "theme" on the impression provoked by tactile sense. We suppose that our image caused by tactile sense emerges from the temporal relationship; we always compare the tactile sense in the past and at the present. If we touch something hard, and then touch something harder, we regard the former as soft. So, the image will be determined by comparison of what/how we touch in the past and at the present; hence we can generate the tactile sense created by mothers' hands by pairing such as hardness and softness and we can generate a rhythm of tactile sense by designing the pair of tactile senses.

1.13 Aesthetical Principle of Composing Tactile Sense

We have analyzed the grammar of massages, we asked the inventor of a system of massage method, Face Therapie method, to break down massages into 42 basic techniques and examined their tactile impressions by using the semantic differential method (SD method) and Principle Component Analysis, we found that the basic techniques are divided into 6 categories (Fig. 1.17).

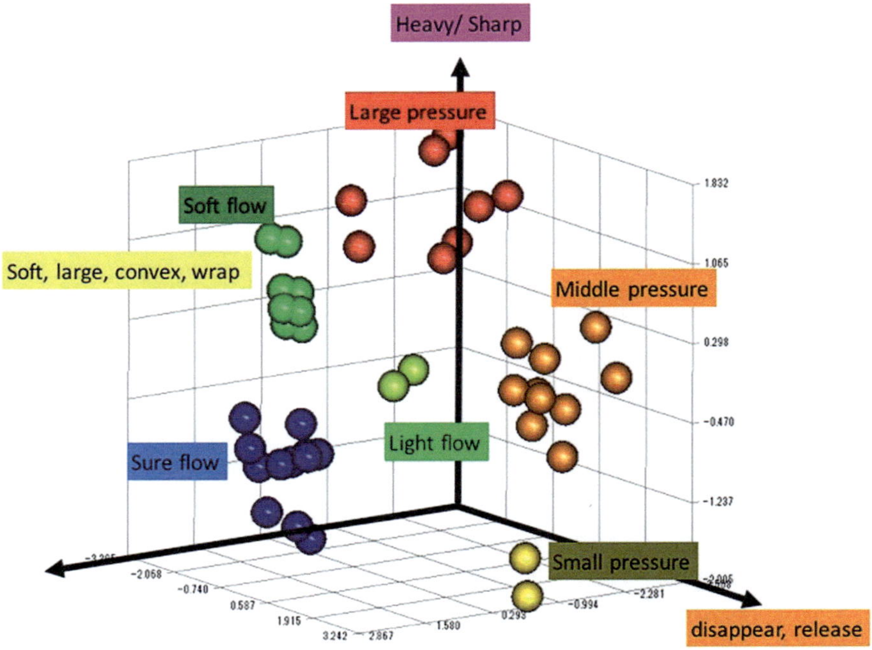

Fig. 1.17 Impressions analysis of basic massage techniques by using semantic differential method and principle component analyses

Based on the six basic massages clusters. We analyze transition rules among these clusters. By using Tactile Score, we denote the massage, which has been selected by over 200,000 customers for more than twenty years and extract transition rules among the clusters and we found out that basically the massage is composed of swapping between two main categories, that are "Light flow" and "Sure flow", where massages of "Light flow" clusters are characterized by "simple and inorganic", while "Sure flow" clusters are "complex and organic"; and in some cases swapping transitions go through very small ("soft flow" tactilely means very small pressure), small, middle, and large pressures, however the transitions do not go through among them, (Fig. 1.18).

From this result, we conjecture that the composing principle of Rational Tactile Score, which makes something aesthetic is as follows;

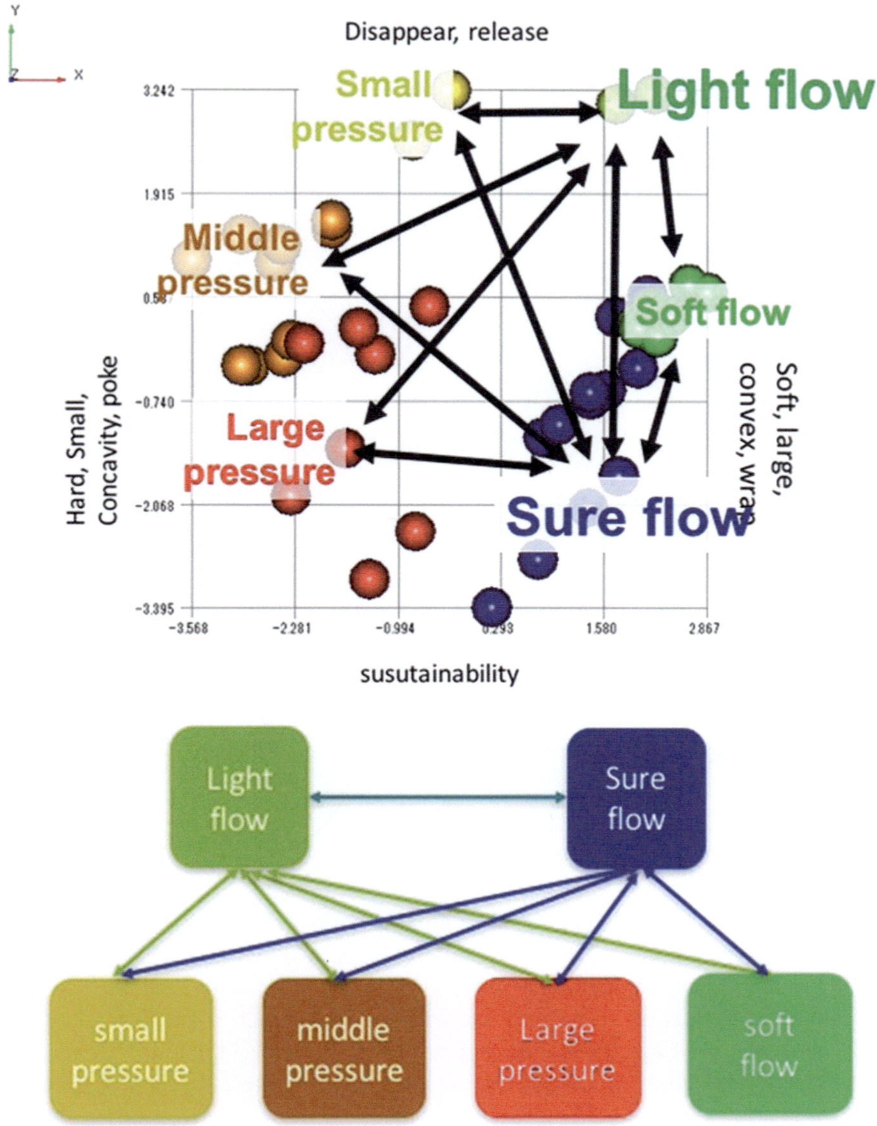

Fig. 1.18 Transitions among clusters, obtained from analyzing a popular massage by using tactile score; we denote the massage by using tactile score and break into basic massages. "soft flow" (below, right) denotes very small pressure tactilely

1.14 Conjecture: Composing Principle of Rational Tactile Sense

Composition of Tactile Scores is rational if and only if compositions of transition arrows on the basic massages: $id \circ H$, $G \circ F$ and $H^{-1} \circ id^{-1}$, $F^{-1} \circ G^{-1}$ are respectively commutative;

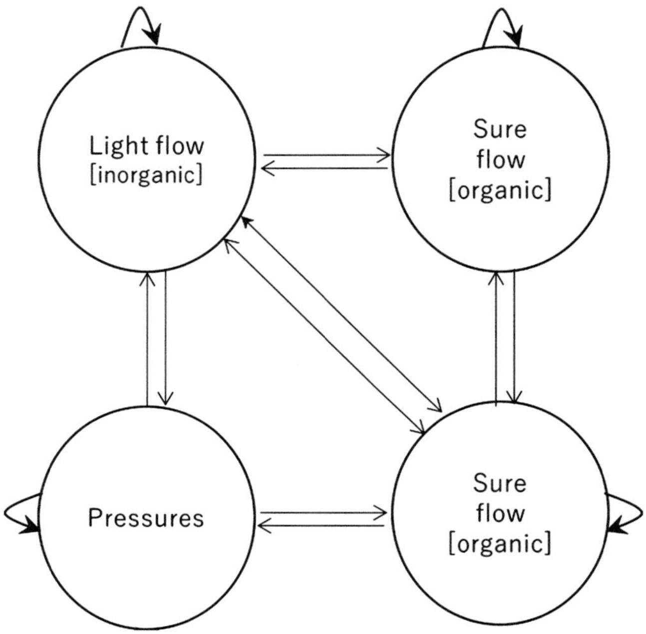

where arrows, H, F, id, H^{-1}, F^{-1} and id^{-1} are transition and inverse transition arrows on non-empty set of Tactile Scores describing basic massage.

It is noticed that the arrows need not to be maps on sets and each unit of basic Tactile Scores needs not to be a set, we will call it as object; if an arrow f is given then objects A, B emerge as $A \to B$, in general empty objects in A and B are allowed but in this contribution, all objects are not empty. We will call object A as Domain, B, Codomain.

Also, we stress that the principle of Tactile Score composing is the method of arrows composing, not basic massages (objects); Tactile Composing is totally different from a "Jigsaw puzzle", objects are not pieces of the puzzle and the "shape of pieces" are changing through composing. If an arrow from "Light Flow, Organic" to "Small pressure", which illustrates "Organic small pressure" massage, while the arrow from "Sure Flow" to "Small pressure" is "Inorganic small pressure" massage and they are different. A composing arrow composes Tactile Score, and the arrows are principle.

Readers may think, the diagram of conjecture is redundant because of omitting id and id-1, we can transform this tetragon diagram into a triangle, which means that arrows are commutative (exchangeable). However, it is not commutative; according to the analysis of massage, the set of Pressures includes four different arrows f_i $(i = 1, 2, 3, 4)$ and their inverse arrows;

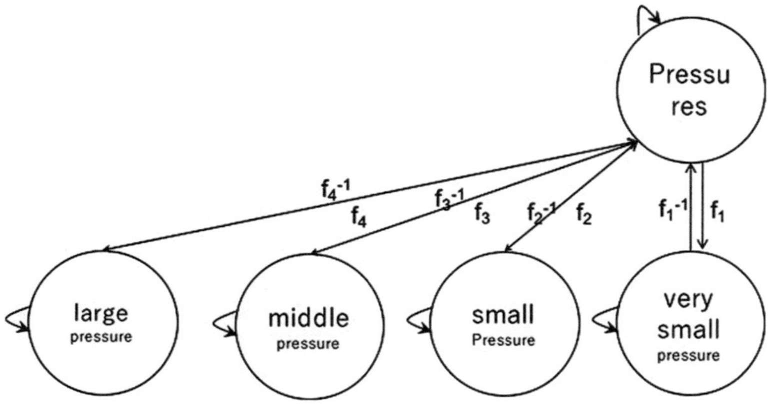

Hence

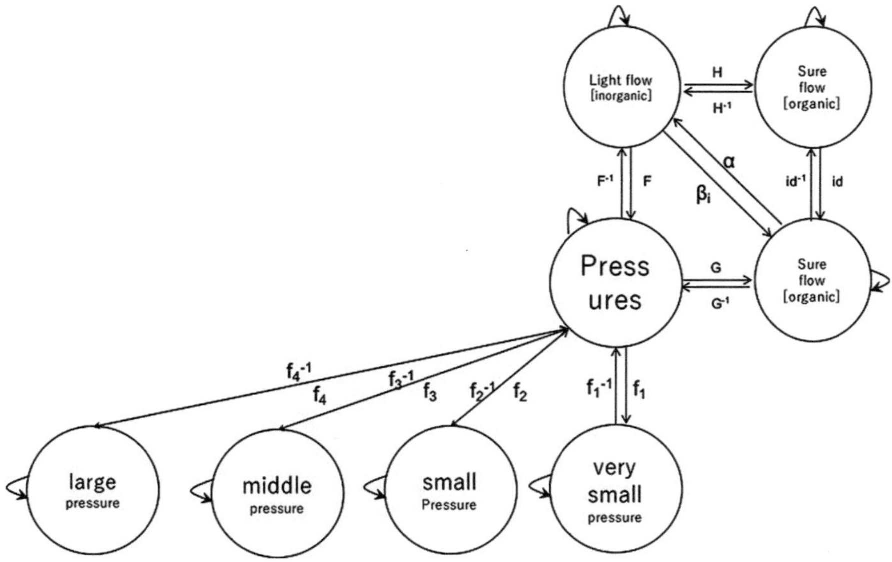

it means that;

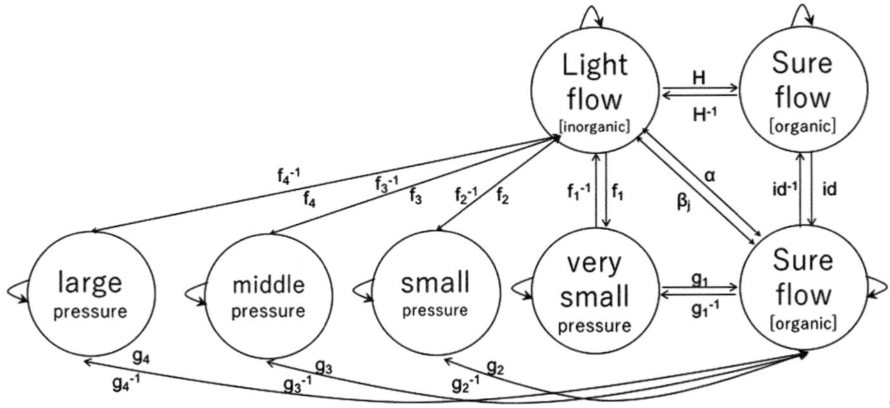

so, composition arrows such as $\beta_i = g_i \circ f_i \neq g_k \circ f_k = \beta_k$, $(i \neq k)$ are different, while the arrowαis unique. Therefore, in order to satisfy the condition of the Conjecture, a transformation $\mathcal{F}(f_i), \mathcal{F}(f_i^{-1}) = f, f^{-1}, \mathcal{F}(g_k), \mathcal{F}(g_k^{-1}) = g, g^{-1}, (i,k = 1,2,3,4)$,which transforms non-commutative diagram to commutative is required;

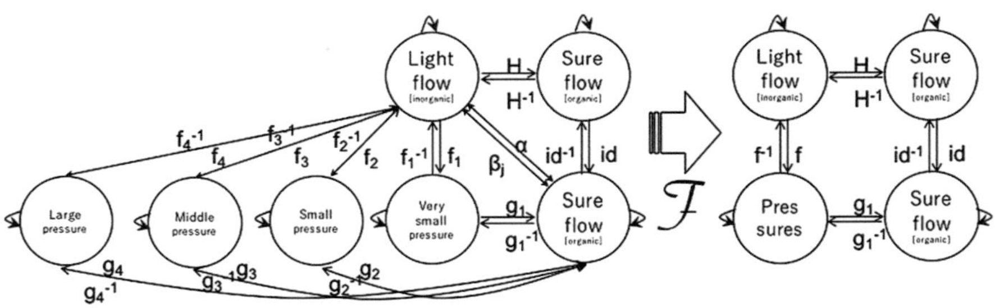

Although it seems that $\mathcal{F}$ is defined only for arrows, the definition of an arrow includes definition of Domain and Codomain; if arrows are changed their Domain (objects of left hand of an arrow) and Codomain (right hand of an arrow) are also changed; because Domain and Codomain are given by an arrow, so if an arrow is changed, then automatically objects will be changed.

Attributes of Tactile Score, arrows in the diagram, are denoted by pressure, contact area and time duration; we call them as "Pressure, P", "Square, S" and "Velocity, V" so practically an arrow is defined as $\varphi(P, S, V) \equiv f_i$, g_k (and their inverse arrows) and the transformation $\mathcal{F}$ transforms from arrows to arrows between the objects of $\varphi(P, S, V) \subseteq H$, vice versa for inverse arrows.

In the transformation $\mathcal{F}(\varphi)$, all Codomains of $\mathcal{F}(f_i)$ are "Pressure" and Codomains of $\mathcal{F}(g_k)$ are "Sure flow" (vice versa for inverse arrows); so $\mathcal{F}(f_1) = \mathcal{F}(f_2) = \mathcal{F}(f_3) = \mathcal{F}(f_4) = f \subseteq H$ and $\mathcal{F}(g_1) = \mathcal{F}(g_2) = \mathcal{F}(g_3) = \mathcal{F}(g_4) = g \subseteq H$, vice versa for inverse arrows. Therefore, by $\mathcal{F}$ the diagram becomes commutative and condition of Conjecture is satisfied.

Fig. 1.19 Example of rational tactile score

Fig. 1.20 Irrational tactile score

Practically, we can give φ such as a constant arrow (constant morphism), $\varphi(p, s, v) \equiv p \times s \times v = c \subseteq H$; hence we can derive the relational expression of Rational Tactile Score, RTS as

$$P \times S \times V = \text{Constant},$$

where P denotes pressure, S, contact area and V, velocity of touching.

1.15 Verification of Relational Expression of Rational Tactile Score, RTS

Through investigation of massages, we are able to obtain the relational expression of principle for composing Rational Tactile Scores, RTS; in order to verify effectiveness of RTS, we compare rational Tactile Scores (Fig. 1.19) with irrational Tactile Scores (Fig. 1.20). We examine beauty effects (tighten and lift face skin and lighten skin tone), Stress marker and Brain activation and we confirmed rational massages increase beauty effects, decrease stress marker and give affection to brain activation; we call the massage, by rational Tactile Score as "rational massage", by irrational Tactile Score, "irrational massage".

Let us make sure that this example of rational Tactile Sore satisfies RTS; since no special symbols denoting contact areas, contact areas do not change and we can ignore S and the relational expression is $P \times V = \mathit{Constant}$; where the first tactile note is the half note with strong pressure and the second note is lighter than I, from $V = \mathit{Constant}/P$, *V*should be faster than before (quarter note); and because 4th note is heavier than 3rd note, V should be slow (half note). And by breaking RTS, we can obtain an irrational Tactile Score; readers will be able to see why it is an irrational Tactile Score;

1.15.1 Beauty Effects

RTS has been obtained through investigation of massages (*Face Therapy,* TM) in beauty salons; especially, we picked up the most popular massage to investigate, which has been selected by more than 200,000 customers for approximately 20 years; hence its beauty effects of the massage have been proved by "social experiments" and the problem is whether we can abstract the essence of composing massage. So, we compose a rational Tactile Score, which is different from investigated one and give hands massage by it.

In order to compare before and after massage, we take photo images of face before and after the massage and compare the changes of the face contour and lightness of skin tone by using image processing. It showed that the contour of the face tended to be lifted up, especially around cheeks and the brightness of the tone of skin tended to be lighten (Fig. 1.21).

After we proposed Tactile Score, it has been used for training of therapists in beauty colleges and it has proved that Tactile Score can deliver the massage techniques; this result confirms these conventional experiences.

1.15.2 Stress Mitigation

We investigate the effect of massage on stress relief; by examining Amylase concentration in saliva before and after massage. It has well known that Amylase concentration indicates the degree of mental stress; low concentration indicates the stress is weak, *vice versa*. We compare rational massage with irrational massage. It has confirmed that Amylase concentration decreases by rational massages, while irrational massage increases it. This result shows that the stress response systems are able to distinguish rational massages from irrational massages (Fig. 1.22).

1.15.3 Near Infrared Spectroscopy, NIRS

We investigate the effect of massage on brain activation by using Near Infrared Spectroscopy, NIRS ; we compare the brain activations of rational massages with irrational massages. For this experiment, we compose rational and irrational Tactile Scores and transform them into sound vibrations by using the automatic conversion system from a Tactile Score to a sound vibration, which we have developed (Usuzawa and Suzuki 2017); and we prepare a sequence of sound vibrations as; Pause (5 sec.): Rational (5 sec.): Pause (5 sec.): Irrational (5 sec.): Pause (5 sec.) and it was repeatedly given 10 times for each subject through a vibrator on the dimple of palm. Except for examining brain activities, before and after experiment (giving

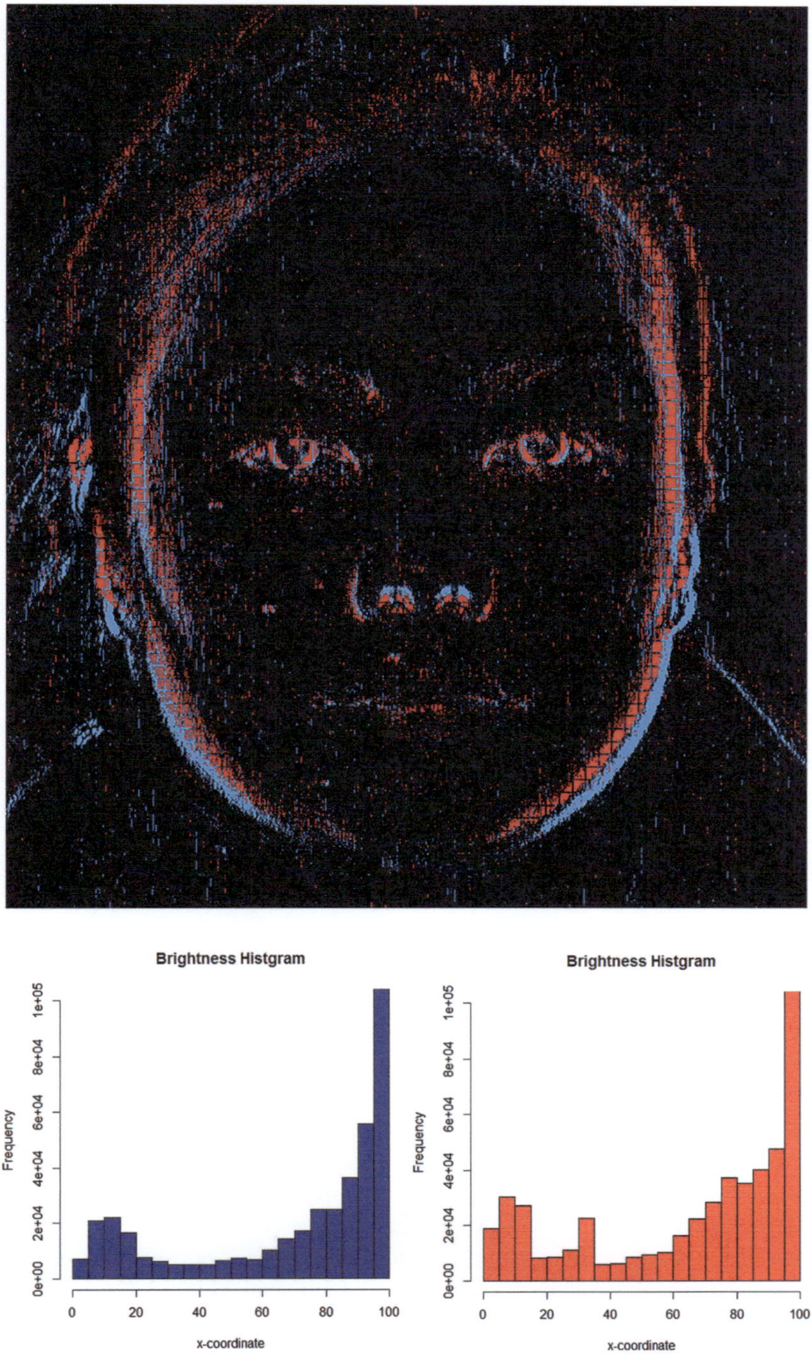

Fig. 1.21 Blue line illustrates the shapes of a face before the massage, and red line, after the massage; histograms illustrate brightness of the facial image before (67.06 average of three subjects) and after (73.21) the massage, total value of brightness increases

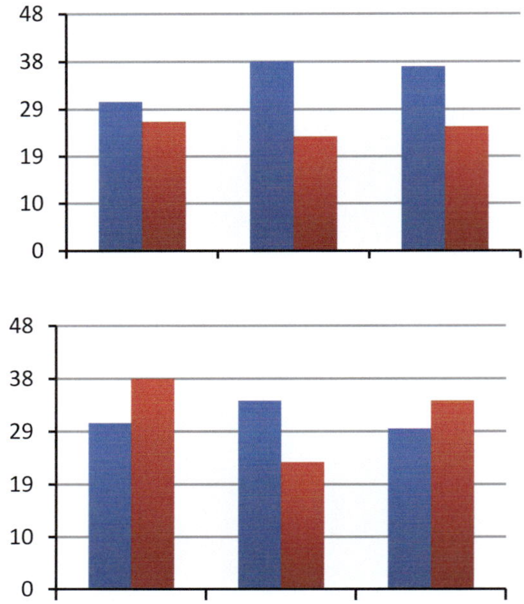

Fig. 1.22 The result of the experiment for analyzing salivary alpha amylase when subjects receive massages that are rational (above) and irrational (below); the number of subjects are three, blue bars illustrates before massage and red, after. The vertical axis illustrates the score of salivary alpha amylase; if the value 0 to 30 (the degree of stress is small), 40-45 (fair), 46-60 (large) and 60- (huge)

Table 1.3 Results of experiments; we took 29 subjects for the experiments, the average of age was 22; subject can be divided into two groups their responses for Rational vibration are higher than Irrational vibration, *vice versa*; the reason why the total of subjects of Questionnaire is 1 less than others, one subject could not tell which of vibrations was comfortable.

	Rational dominance	Irrational dominance
Brain activities (NIRS)	10	7
Amylase	9	8
Questionnaire	10	6
Sum	29	21

vibrations) the concentration of Amylase in saliva were examined and each subject asked to answer the questionnaire of which vibration was comfortable.

We took 29 subjects for the experiments, the average of age was 22; we divided subjects into two groups; they are a group that activates brain activity with Rational vibration and a group that Irrational vibration activates; we will call the groups as "Rational dominance" and "Irrational dominance". In addition, we grouped even amylase concentration response. They are a group in which Rational vibration lowers amylase concentration and a group in which Irrational vibration lowers it. The number of Rational dominance is larger, however there are no statistical difference in between Rational dominance and Irrational dominance [Table 1.3].

Interestingly, it was confirmed that the preference in the questionnaire and the response of brain activity and amylase concentration may not match. For example, a subject answered Irrational vibration as comfortable as the questionnaire, but Rational vibration activated the brain and lowered the amylase concentration. The

Table 1.4 The number of subjects who matched the preference of questionnaires, brain activity, and amylase concentration, respectively.

	Rational dominance	Irrational dominance
Brain activities (NIRS)	7	3
Amylase	7	5
Sum	14	8

coincidence rate between the questionnaire result and brain activities is 62.5 percent, Amylase concentrations, 75 percent. And the number of subjects who matched the preference of questionnaires, brain activity, and amylase concentration all together in Rational dominance is 6/29 (20.0 percent), Irrational dominance is 3/21 (14.3 percent). It shows that aesthetic preferences (comfortableness) in consciousness level are different from biological responses.Hence, we selected subjects whose preference in the questionnaire was consistent with brain activity and amylase concentration (Table 1.4); and we confirmed that brain activity and amylase concentration have statistically significant differences respectively in between Rational and Irrational dominance.

If the preference of the questionnaire is consistent with the decrease in amylase concentration, it tends to be linked with brain activity (Fig. 1.23 above); no matter how it is rational or irrational vibration, higher preference vibration reduces the concentration of amylase and activates the brain. On the other hand, when the preference of the questionnaire is consistent with the brain activity (Fig. 1.23), regardless of preference, Rational vibration lowers amylase concentration.

It has shown that in 3/4 cases, when brain activities are high, Amylase concentration is lowed, which indicates that relaxation through tactile sense, that is aesthetics of tactile sense, is closely related to brain activities. So at least on tactile sense, sensory stimulations evoking brain activities would be principle of aesthetics of tactile senses.

Only the exception is in the Irrational dominance when preference of the questionnaire is consistent with the brain activation; where the relation between Amylase concentration decreasing and brain activities are reversed; when brain activities are low, Amylase concentrations are also lowed, and high brain activities increase Amylase concentrations; it would suggest for future investigation that there is another way to bring relaxation.

1.16 From Designed Tactile Score to Visual, Audio and Haptic Sense

We have proposed the tactile score for creating tactile sense produced by massage; we have investigated how massages are composed and found out there exists a "grammar" and confirmed the effectiveness for beauty treatment by "rational" massages.

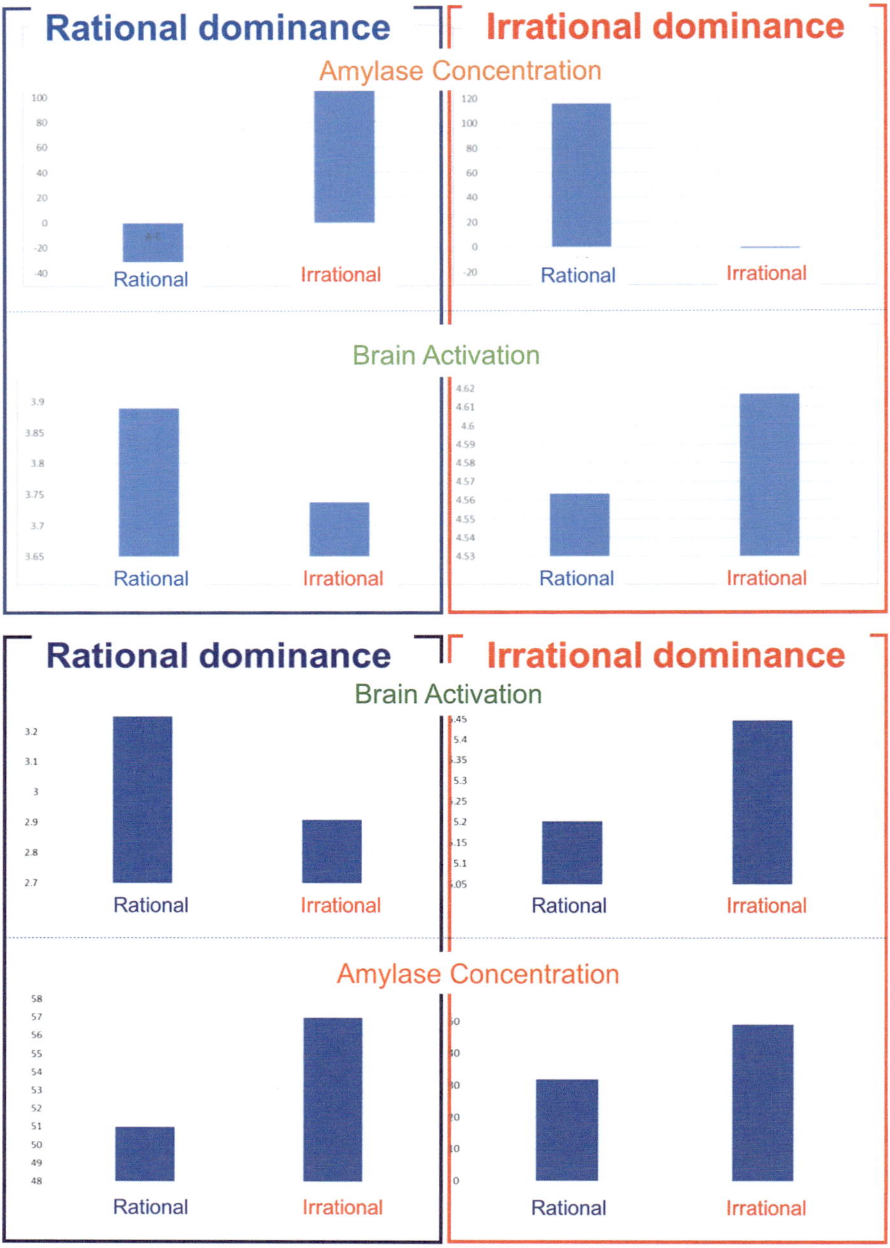

Fig. 1.23 Subjects whose preference in the questionnaire was consistent with brain activity and amylase concentration; above: preference of the questionnaire is consistent with the decrease in amylase concentration, below: preference of the questionnaire is consistent with the brain activation

Based on these results, we have attempted to utilize it as a sensory language not only for composing massages; we extract tactile sense from texts and translate it into multimodal media (dance notation, computer graphics, sound vibrations and so on). Tactile scores can use as a sensory language, which is able to extract tactile sense from multimodal media and describe it as "intermediate forms" by using the tactile score; we are able to analyze, edit and compose the intermediate forms and re-translate into multimodal media again. Through this media translation, we will be able to name the things that we have been named as quality without name.

1.17 Final Remarks

If the computation is Natural Computing, which has been performed itself and does use a device for computing outside, it requires Aesthetics to halt computing. Hence Aesthetics is not only for Arts or Philosophy but also computation in nature, which has been considering in Physics, Chemistry, Biology, etc., where the Aesthetics has been named different terms as "stable", "robust", "potential" and so on. Since Computer Science raised in the middle of twenty centuries and had not grown as a natural science, almost all the scientists have not recognized that Algorithm is a media for describing and considering nature as equations but just an instruction for advanced electric abacus. As a founder of Natural Computing, Leibniz considered, Algorithm is a method of considering nature. And in the Natural Computing, Aesthetics is the foundation.

In this contribution, we showed our attempts of Computational Aesthetics: in order to investigate Aesthetics of visual art, we created various art works and try to extract people's visual palatability; and found that visual artworks in between order and chaos have high palatability, which has stated by Paul Klee (1956). This result would be applicable to giving creativity in Artificial Intelligence.

Then we investigate Aesthetics of tactile sense based on massages; we proposed method of describing tactile senses, Tactile Score and extract the Algorithm of creating aesthetic state by tactile sense. And we verify its effectiveness through examining biological responses of human. It would be applicable to design biological condition algorithmically. Based on these preliminary results, we will expand Computational Aesthetics in molecular, plants, insects, animal and social communications.

Appendix: Method for Evaluating Output Pictures

Questionnaire Using Semantic Differential Method

We classified output pictures by generations such as early, middle, and late periods, and then we did a questionnaire using Semantic Differential, SD method (Osgood, 1957) for researching what evaluation of output pictures were given in which generation. SD method is one of methods for evaluating sensibility aspects and it is used in many fields.

The feature of SD method is to evaluate a certain stimulating concept with a pair of adjectives. First, we draw up a questionnaire with pairs of adjectives that describe contrasting words and set a scale of 1 to 5 or 7 among a pair of adjectives. We evaluate a concept with respect to each pair of adjectives, and then we can compare the differences among stimuli. This method has been used for psychological experiment since early times, and deserves a long-established method. We used the studies by Tsutsui and Goldberg (2001) and Inoue et al. (1994) as a reference for pairs of adjectives.

Factor Analysis Method

Factor analysis is a method for specifying some elements that affect a number of observed data. This element is called a factor. In this study, we regarded the questionnaire items using SD method as observed data, and used this method for searching factors that affected those results. We conducted the factor analysis using Excel 2010 (Social Research Information Co., Ltd.).

Correlation Coefficient

We can examine the strength of relationship between two items with correlation coefficient. If the correlation coefficient is positive, the item has positive correlation, and if it is negative, the item has negative correlation. The survey of the correlation coefficient of each item suggests how the items relate to each other. It is called a correlation matrix whose correlation coefficient is calculated from each item and the coefficient is described in a matrix.

Identification of the Number of Factors

We conduct the factor analysis with the correlation matrix and determine the number of factors. The following standards are generally used for determining the number of factors.

1. Gutman standard: Employ a factor whose eigenvalue is more than 1.
2. Scree standard: Plot the size of the eigenvalue and observe the change, and then extract the factor.
3. Employ the factors those of cumulative contribution ratios are more than approximately 60 %.
4. Employ a factorial structure with possible meanings.

In this study, we decided the number of factors as 2 factors that had high contribution ratios to interpret the results easily.

Factor Nomenclature

We conduct a factor analysis with the number of factors determined in the previous section. We denominate the factor by reference to the factor loading of the questionnaire item that is affected by the factor.

Multiple Comparison Test

Multiple comparison test is used for comparing each data group without raising the significance level when there are more than 3 data groups. In this study, we compared each item in early, middle, and later generations and examined the significant difference with Tukey-Kramer method.

References

Akiba F Private communication
Choi I Private communication
Inoue Y, Shimamoto T, Sakamoto A (1994) A module placemnet using genetic algorityms. J Inst Electron Inf Commun Eng 77(8):1189–1181
Klee P (1956) Das bildnerische Denken. Benno Schwabe & Co. Verlag, Basel
Kurzweil R (2005) The singularity is near. Viking Books, New York
Menninger KW (1969) Number words and number symbols: a cultural history of numbers. MIT Press
Okamoto T (1954) Art today. (in Japanese "kon nichino geijyutsu"). Kobunsya
Suzuki R Private communication

Suzuki Y Foundation of natural computing (in Japanese "sizen keisan no kiso"), kindai-kagakusya, to be published
Suzuki T, Suzuki Y (2014a) Modeling and evaluating the process of creating paintings with evolutionary computing. Proc Artif Life 14:572–579
Suzuki Y, Suzuki R (2014b) Tactile score: a knowledge media for tactile sense. Springer Verlag
Tsutsui S, Goldberg DE (2001) Search space boundary extension method in real-coded genetic algorithms. Inf Sci 133(3–4):229–247
Usuzawa A, Suzuki Y (2017) Analysis of relationship between vibrotactile sensation and brain activity by near infrared spectroscopy, Graduation Thesis of School of Informatics, Nagoya University,
Yanaihara I (2010) Giacometti note., Unexpurgated edition (in Japanese "kan pon Giacometti techou"), vol I, II. Misuzu Syobou

Chapter 2
On the Possibility of Computational Aesthetics of Massaging

Fuminori Akiba and Yasuhiro Suzuki

Abstract Some researchers of philosophical aesthetics object to our attempt of the aesthetics of massaging. In this paper, however, we briefly show that an aesthetics of massaging is possible. We take up some literatures that are often referenced for the purpose of criticizing our attempt and point out their misunderstandings and then make it clear that these literatures do not treat tactile sense itself. Finally, we show both the current results of computational aesthetics in massage and expectations for the near future.

Keywords Computational aesthetics · Tactile sense · Tactile score · Massage · Translatability of senses

2.1 Introduction: Aesthetics of Massaging?

Some researchers of philosophical aesthetics object to our attempt of the aesthetics of massaging. They cite two opposing reasons. Some say that the concept of the aesthetics of massaging is impossible. Others say that we already have enough aesthetics of touch, so there is no need for a new aesthetics of massaging.

In this chapter, however, we try to demonstrate that an aesthetics of massaging is possible. Since we received objections from anonymous reviewers, we cannot reply directly to them. Therefore, in the following sections we take up some literatures to which the reviewers referred and briefly point out their misunderstandings and make clear the fact that these literatures do not treat tactile sense itself. Finally, we add some results of computational aesthetics of massaging and explain why computational aesthetics is suitable for investigating massaging.

F. Akiba (✉)
Graduate School of Information Science, Nagoya University, Nagoya, Japan
e-mail: akibaf@is.nagoya-u.ac.jp

Y. Suzuki
Department of Complex Systems Science, Graduate School of Informatics,
Nagoya University, Nagoya, Japan
e-mail: ysuzuki@nagoya-u.jp

Y. Suzuki et al., *Computational Aesthetics*, SpringerBriefs in Applied Sciences and Technology, https://doi.org/10.1007/978-4-431-56844-5_2

2.2 Objections and Replies (1)

2.2.1 Kant's Criticism of Massaging

The first objection is that an aesthetics of massaging is impossible. We quote Kant's criticism of massaging from his well-known, "Kritik der Urteilskraft."

> [T]he voluptuaries of the Orient find so comforting when they have their bodies as it were kneaded, and all their muscles and joints softly pressed and flexed [...] the moving principle [...] is entirely outside us. (Kant 2000: 155)

Some researchers seize upon this passage and insist that an 'aesthetics of massaging' is impossible. But we think they make a mistake because they forget the fact that the passage is located in a discussion on the 'sublime.'

2.2.2 The Passage Is Located in a Discussion on the 'Sublime'

According to Kant, 'sublime' is the term that should be assigned not to natural phenomenon but to the mental attitude of human beings. For example, a huge, violent hurricane causes fear, a strong disturbance [=a movement of the mind] in us. However, against such a disturbance, we human beings dare to try to make its mechanism clear and search for a way we can reduce the damage from this hurricane to the minimum. In this effort we feel intellectual pleasure and this pleasure entirely results from our inner intellectual principle: we must overcome the fear of a huge hurricane. Therefore, we—not the hurricane—are sublime. This is a story of enlightenment typical of the 18th century.

In this context, Kant criticizes the voluptuaries of the Orient because the moving principle is entirely outside us, and he says that their attitude does not deserve the name 'sublime.' Massaging causes some movements in us, but it only moves us from the outside and has nothing to do with our inner intellectuality. Kant's argument is based on the story of enlightenment.

2.2.3 The Story of Enlightenment Cannot Forbid Us to Conduct Research into Massaging

Certainly this story of enlightenment is important for Kant's program of aesthetics. According to Kant, the beauty and the sublime must facilitate our understanding of nature and at the same time it must facilitate our moral feelings (Akiba 2013: 125). And this idea is still important for our aesthetics. However it cannot forbid us to conduct research into massaging because we cannot decide whether massaging has some inner principles before we really investigate it. As we show in the next section, no aesthetics except ours really investigates massaging or tactile sense itself.

In addition, our research target is not the bodily reaction of the person who receives massage but the way of massaging from the viewpoint of the massager. It has nothing to do with the receiver's body being moved by the massaging. Therefore, we say again, even if the story of enlightenment is important for the program of aesthetics, it cannot forbid us from conducting research into massaging.

2.3 Objections and Replies (2)

The second objection is that there are already various kinds of aesthetics of touch, so there is no need for a new aesthetics of massaging at all.

Certainly, one can at least think of J.G. Herder (1744–1803), Herbert Read (1893–1968), and Gilles Deleuze (1925–1995) as some of the representatives who have tried to theorize touch. However, their interests are not in tactile sense itself. Rather, they pay much attention to an original unity of the senses (common sense) or to a traverse among the five senses. 'Feeling' or 'the haptic' are the names they give it, respectively.

2.3.1 Herder Does Not Treat Tactile Senses Itself

Herder discriminates between two different touches. One is 'tactile sense' (in German; Tastsinn), the other is 'feeling' (in German; Gefuehl). Tactile sense is one of the five senses and these five senses are based on feeling. Feeling is the most basic sense, a kind of common sense, and Herder places great importance on it. He does not treat tactile sense itself (Sugiyama 2006: 6 and n.6). In his analysis, there appear to be at least four varieties of touch; (1) tactile, (2) haptic, (3) proprioception, and (4) kinaesthetic perception (Zuckert 2009: 291). However, what we really want to know is the principle that makes such variety possible.

2.3.2 Sculpture Is Not the Art of Tactile Sense

Herder thinks that the art of feeling is sculpture (Herder 1994: 257). Like Herder, Herbert Read, one of the most famous art critics of the 20th century, also regards sculptures as the art of touch. He calls it the "art of palpation."

> For the sculpture, tactile values are not illusions to be created on a two-dimensional plane: they constitute a reality to be conveyed directly, as existent mass. Sculpture is an art of palpation—an art that gives satisfaction in the touching and handling of objects. That, indeed, is the only way in which we can have direct sensation of the three-dimensional shape of an object (Read 1977: 49).

However, the word 'palpation' is only a metaphor and, in fact, it means a cooperative, complex process of touch and vision. Read's aesthetics of sculpture cannot directly help the aesthetics of tactile sense.

2.3.3 *The Researcher Who Emphasizes the Traverse among the Five Senses in a Haptic Eye, Refuses the Research of Tactile Sense Itself*

Another philosopher often referred to is Gilles Deleuze. But his topic is the paintings of Francis Bacon and Deleuze analyses it from the viewpoint of a traverse. Here we understand the word 'traverse' as encroachment among the senses. For example, the visual inevitably evokes the tactile or the audible evokes the visual and so on.

> The painter (Francis Bacon) makes visible a kind of original unity of the senses. This operation is possible only if the sensation of a particular domain is in direct contact with a vital power that exceeds every domain and traverses them all. (Deleuze 1981: 37)

Deleuze says that this traverse needs "a haptic eye, a haptic vision of eye" (128).

Everyone can easily understand Deleuze does not treat tactile sense itself. From what we showed above, we can safely say that these are not aesthetics of tactile sense. Objections to our attempt to an aesthetics of tactile sense are completely unreasonable.

Of course, we admit that the traverse among senses is important, and we do not think we can completely isolate tactile sense from other senses. However, if we really want to understand the traverse, at first we must construct a simple model of tactile sense and then consider how it can interact with other senses. This is our opinion.

2.4 Current Results of Computational Aesthetics of Tactile Sense

In this final section, we introduce a brief history of our research and its results. In addition, we briefly explain why 'computational' aesthetics has an advantage in investigating massaging.

2.4.1 *Massage as a Research Object*

Aesthetics is a general science. It needs results not only from philosophy and ethics, but also from various fields of science, for examples, biology, cognitive science, social psychology, and information science. But as we saw above, there is no aesthetics of tactile sense. We had to start from zero. So we chose massage as the research object because no one can deny the fact that massage simply concerns tactile sense.

2.4.2 Problems and the First Result: Invention of Tactile Score (TS)

However there were problems in investigating massage. One problem was that the effect of massage seemed to vary from person to person. It seemed to depend on ages, gender, and the cultural background of the person who received massage. But we could easily avoid this problem because we decided to start our research from the viewpoint of the massager. What we would like to know is not the varieties but a few creative principles that produce such varieties. This decision is based on our idea of 'harness' (Suzuki 2013). We do not have to know everything directly about a system. If we can find a few accessible principles of the system, we can indirectly control the system (cf. Akiba 2013: 126–127). We focused only on the effective massages in the aesthetic salon (TFT) and decided to analyze the movement of the massaging.

The more serious problem was that there was no means to record massage. It quickly disappears as the massager moves her hands. In order to solve this problem we invented 'tactile score (R).' Based on our experience we had found that there are three elements in massaging: contact area, speed, and pressure. We assign these three elements to a five-line score. This is the tactile score (Suzuki and Suzuki 2014: 21–26). This made it possible for us to record and visualize massaging. With the tactile score, we formalized the effective massages and analyzed them by using principle component analysis. As the result, we made it clear that the first principal component is the characteristics of touch (soft-large, blow-wrap), and the second is the time variation of touch (disappearing-releasing), and the third is the pressure (heavy-sharp) (Suzuki and Suzuki 2014: 27–29).

2.4.3 Why Is 'Computational' Aesthetics Suitable?

Finally, we explain why 'computational' aesthetics is suitable for investigating massaging. The reason is quite obvious. The important things in computing are movement and timing. And movement and timing are also important for massaging. In this sense massaging is a kind of computing. This is why 'computational' aesthetics is suitable for an investigation of massaging. And a tactile score, like a musical score, is good at displaying movement and timing.

2.4.4 *Expectations for the Near Future: Translatability of Sense*

By using tactile score, we are now investigating translatability among the senses. First, we construct a model of massaging and then translate it into music, image, and text. At the same time, we translate music and text into massaging (Suzuki et al. 2014). Then if we find a significant relationship between them, we could empirically say something about the traverse among the five senses and reconnect our aesthetics with the aesthetics we referred to in the previous sections.

Acknowledgements This work was supported by JSPS KAKENHI No.24520106 and No. 23300317

References

Akiba F (2013) The significance of natural computing for considering computational aesthetics of nature. In: Natural computing and beyond, proceedings in information and communications technology, vol 6. Springer, pp 119–129

Deleuze G (1981) Francis Bacon: the logic of sensation. University of Minnesota Press

Herder JG (1994) Plastik, Werke, vol Bd. 4. Deutscher Klassiker Verlag

Kant I (2000) Critique of the power of judgment, Cambridge

Read H (1977) The art of sculpture, Bolligen Series. Princeton

Sugiyama T (2006) Herder's theory of sensorium commune: The birth of the concept of synesthesia, in BIGAKU (aesthetics). Jpn Soc Aesth 57(1):1–14

Suzuki Y (2013) Harness the nature for computation, natural computing and beyond. Proc Inf Commun Technol 6:49–70. http://link.springer.com/chapter/10.1007%2F978-4-431-54394-7_5

Suzuki Y, Suzuki R (2014) Tactile score: a knowledge media for tactile sense. Springer

Suzuki Y, Watanabe J, Suzuki R (2014) Translation from text to touch: touching a "Japanese old tale". In: Proceeding CHI EA '14 CHI '14 extended abstracts on human factors in computing systems, pp 685–694

Zuckert R (2009) Sculpture and touch: Herder's aesthetics of sculpture. J Aesth Art Crit 67(3):285–299

Chapter 3
Herder on "Sentio, Ergo Sum": Seen from His Remarks on the Color Harpsichord

Takashi Sugiyama

Abstract This chapter outlines the unique aesthetics of German philosopher Johann Gottfried Herder (1744–1803) by paying attention to his remarks on the "color harpsichord". "Color harpsichord" is a curious instrument invented by French Jesuit Louis-Bertrand Castel (1688–1757), inspired by the correlation theory between spectral hues and musical scale which English natural scientist Isaac Newton (1642–1726) proposed in *Opticks* (1704). So Castel attempted to visualize this correspondence between spectrum and musical scale by conceiving an instrument that, on striking a key, shows a color corresponding to the musical note. This color harpsichord was discussed heatedly by many great philosophers of the time (Rousseau, Diderot, Goethe etc.). Herder is among them the one who was most interested in (mentioned most times) the color harpsichord, though anytime negatively. He took the failure of the color harpsichord because of lack of doubt or reflection on sight and hearing being modally different and forces an unnatural mode of perception in reverse of the process of human development, from a "sensorium commune" to a being with five differentiated senses. This argument can be summarized, as Herder himself wrote, into the motto "sentio, ergo sum", as opposed to the Cartesian "cogito, ergo sum".

Keywords Johann Gottfried Herder · Louis-Bertrand Castel · Color harpsichord · Sensorium commune

3.1 Introduction

We generally consider ourselves as having five senses, among which the sense of sight and hearing count as "superior," as opposed to the "inferior" smell, taste, and touch. We find one of the oldest pieces of evidence supporting this in Plato: "Have the sight and hearing of men any truth in them […]? And yet if these two physical senses are not accurate or exact, the rest are not likely to be, for they are inferior to

T. Sugiyama (✉)
Kyoto University, Kyoto, Japan
e-mail: sugiyama.takashi.3s@kyoto-u.ac.jp

Y. Suzuki et al., *Computational Aesthetics*, SpringerBriefs in Applied Sciences and Technology, https://doi.org/10.1007/978-4-431-56844-5_3

these" (*Phaedo*, 65B). And Plato showed even the determination that "the beautiful is that which is pleasing through hearing and sight" at one point, but later withdrew, for "the expression through sight and hearing makes both collectively beautiful, but not each individually" (*Hippias Major*, 298A–303D: trans. by Harold North Fowler).

Plato then came to establish the theory of Ideas (of beauty), but it is interesting that he fell into an aporia with the question of whether sight and hearing are capable of beauty either (only) individually or collectively. If he would have known of "synesthesia," a phenomenon in which one type of stimulation evokes the sensation of another and the most typical of which is "color-hearing," then he could have argued otherwise and might not have needed to withdraw the determination above. Is synesthesia, however, a matter only within the "superior" senses, even if restricted to color-hearing, that is, the synesthesia between sight and hearing? No. In this article, I argue so by referring to a curious instrument that appeared in the eighteenth century and to the remarks on it by a philosopher: the color harpsichord invented by Louis Bertrand Castel (1688–1757) and the remarks on it by Johann Gottfried Herder (1744–1803).[1]

3.2 Invention of the Color Harpsichord and Its Spread

Castel was born in Montpellier and entered the Jesuit order in Toulouse in 1703. Having been trained in southern France, he was sent to Paris as a teacher in 1720. From then until his death, he devoted himself to research and education in mathematics and natural philosophy in Paris.[2]

It was *Opticks* (1704) by Isaac Newton (1642–1726) that inspired Castel to invent the color harpsichord. The French translation thereof was published in 1722 and reviewed by Castel (1723). In *Opticks* Newton reported the experiment of analyzing light with prisms and argued that light comprises seven different spectral hues. Moreover, he correlated these hues to the musical scale, i.e., seven different musical notes (First Book, Part 2, Experiment 7).

Castel attempted to visualize this correspondence between spectrum and musical scale by conceiving an instrument that, on striking a key, shows a color corresponding to the musical note. This is the color harpsichord. After 1725, Castel publicized this idea in several journals (Castel 1725, 1735) and then put it into *Optics of the Colors* (*L'optique des couleurs*, 1740), which recounted, however, not *how* to construct a color harpsichord, but only its *theoretical background*, for example, which color corresponded to which tone.[3] Meanwhile, after Castel's friends urged him to

[1] In Sugiyama (2009), I have already argued for Herder as the pioneer of a philosophical concept of synesthesia, but without referring to the color organ.

[2] For Castel's biography, see Schier (1941), Chouillet-Roche (1976), Franssen (1991).

[3] Yet, Castel adopted here not the theory of Newton, but that of Descartes, who in *Dioptrics* (*Dioptrique,* 1637) mentioned blue, yellow, and red as basic (Descartes 1996, VI 92) and correlated them to the root, the third, and the fifth of triad. In order to make more suitable for the organ, he further correlated not seven colors to the diatonic, but twelve to the chromatic scale (Castel 1740, 173).

actually construct a color harpsichord, he finally did so. Supposedly, he made it public to a small audience twice between the end of 1754 and the beginning of 1755. However, no witness to this remains.[4] Castel himself wrote a letter to a British friend reporting this, in which he was, however, proud only of the audience's excitement and did not describe how the organ was constructed or how it worked.[5] Therefore, the color harpsichord became a kind of "phantom" (Caduff 2002).

Nevertheless, with the German composer Georg Philipp Telemann (1681–1767) being central, the idea itself spread quickly throughout Europe. Telemann traveled to Paris in 1737, met Castel and was struck by the idea of the color harpsichord. Upon returning to his hometown of Hamburg, he published a booklet reporting it (Telemann 1739). German natural philosophers responded to him because they had imported, were interested in, and inspired by Newtonian natural philosophy. As early as 1740, Johann Gottlob Krüger (1715–1759) mentioned the color harpsichord in his book (Krüger 1740–1749, §503). The instrument also spread far to the Petersburg Academy of Sciences led by Leonhard Euler (1707–1783), where in 1741 a member Georg Wolfgang Krafft (1701–1754) lectured on Castel's theory. Being present at and perhaps inspired by this lecture, Krüger deepened his reflection on the color harpsichord and eventually contributed a blueprint of it (Fig. 3.1) to the journal of the Royal Prussian Society of Sciences in Berlin (which offered Euler the position of director of the mathematical class in 1741). This is the only remaining blueprint. Though not the one originally invented by Castel, but of Krüger[6], this adequately aids in guessing what it was like to be. In this figure, the rectangular part surrounded by ABDC does not differ from the normal clavichord, but a huge number of lanterns, G, concave mirrors, H, and colored lenses, I/J, are so arranged in the arc-shaped part surrounded by CDFE as to emit different colored lights. These are allowed to be seen or to be shaded by the covers, X, connected by levers to the keys. Therefore, this instrument was constructed to produce notes similar to a normal clavichord, and to simultaneously emit colored light according to the notes played.

It is interesting from an aesthetic point of view that many great thinkers of the time mentioned this instrument, such as Jean-Jacque Rousseau (1712–1778), Denis Diderot (1713–1784), Immanuel Kant (1724–1804), Moses Mendelssohn (1729–

[4]Rousseau said, "I have seen the famous clavichord on which music is supposedly made with colors" (Rousseau 1959–1996, V 419 = 61f.), but without detailed information regarding when, where, and how it was exhibited.

[5]For the first performance (21.12.1755), "I gave an invitation to fifty persons of the house and from abroad, and lighted a hundred wax candles. There was nothing but acclamation and clapping of hands in consequence for the space of half an hour that I played" (Anonymous 1757, 12). For the second performance, "Every one wanted to see, and see again this novelty. [...] The Harpsichord played, and two hundred persons owned that they had never seen any thing more beautiful, or more brilliant, and all concluded that the Harpsichord was executable, possible, true, and even made" (13).

[6]For the difference between them, see Jewanski (1999, 479).

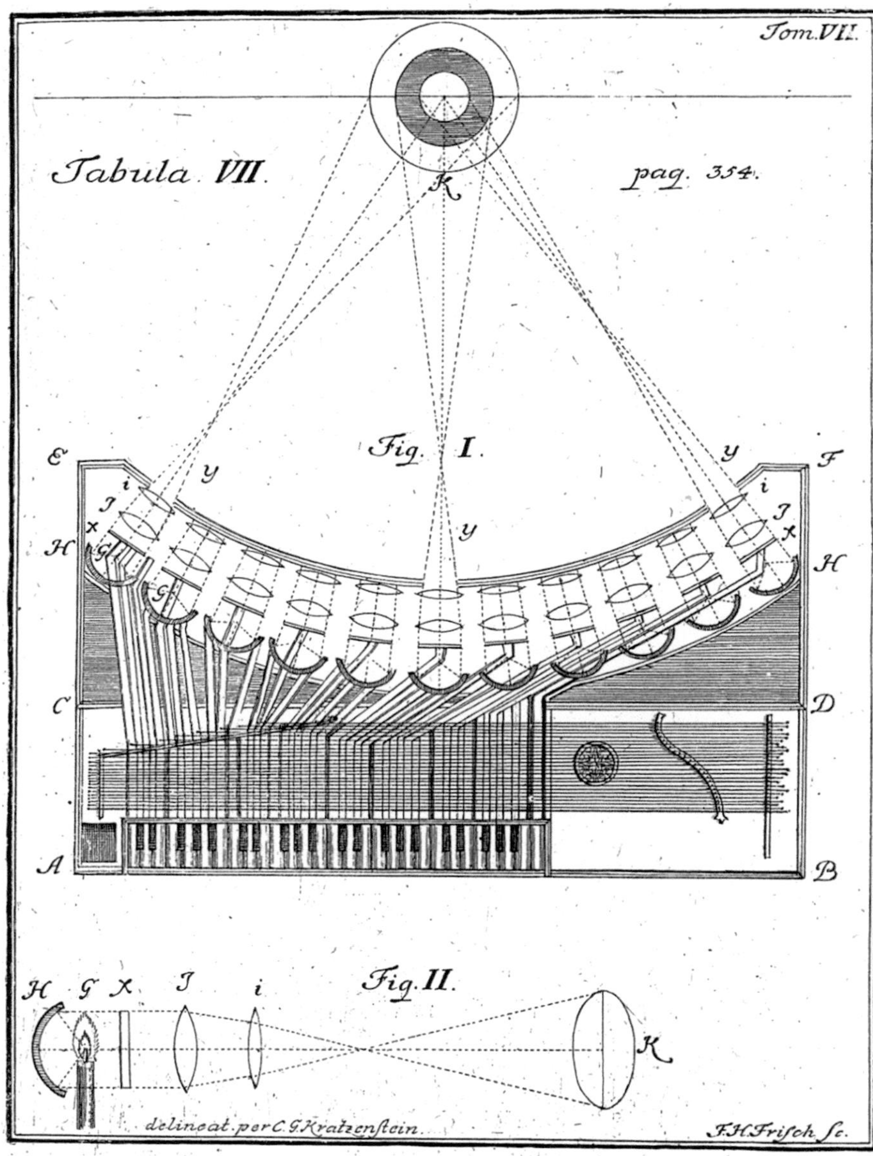

Fig. 3.1. Krüger's Blueprint of Color harpsichord (Krüger 1743, 354)

1786), and Johann Wolfgang von Goethe (1749–1832). This means that the instrument reflects an aspect of the aesthetics of the time, the age of enlightenment. Among these thinkers, Herder is the one who was most interested in (mentioned most times) the color harpsichord.

Table 3.1 Lessing's dichotomy

Painting	Things alongside one another (nebeneinander)	Body (Körper)
Poetry	Things in succession (nacheinander)	Action (Handlung)

3.3 What Do Herder's Remarks on the Color Harpsichord Mean?

The color harpsichord greatly influenced the genesis of Herder's own aesthetics, and especially, his classification of the arts, which was a counterproposal to that of Gotthold Ephraim Lessing (1729–1781). In *Laocoon* (*Laokoon*, 1766), Lessing showed the limits of painting and poetry—subtitle of the book (*die Grenzen der Malerei und Poesie*)—as follows (Table 3.1):

> Objects or parts of objects that exist alongside one another are called bodies. Accordingly, bodies with their visible properties are the true subjects of painting. / Objects or part of objects that exist in succession are called actions. Accordingly, Actions are the true subjects of poetry. (Lessing 1968, IX 94 = Herder 1985–2000, II 192)[7]

This argument was so innovative as to break the artistic tradition "ut pictura poesis." However, in *Critical Forests* (*Kritische Wälder*, 1769; whose First Grove is "dedicated to Mr. Lessing's *Laocoon*"), Herder proposed a counterargument that "the concept of succession is only half of the idea of an action; it must be a succession through force: thus arises an action," because "Father Castel's color clavichord will never deliver actions by playing a succession of colors" (Herder 1985–2000, II 196).[8] He thus held Lessing's dichotomy inadequate (in addition, Lessing said that "under the general term 'Painting,' I desire to be understood the arts of design in all their departments" and hence did not distinguish sculpture from painting [Lessing 1968, IX 5f.]) and instead proposed a following new trichotomy, by means of the concept of "force," in *Sculpture* (*Plastik*, 1778) (Table 3.2):

We have one sense that perceives external things alongside one another, a second that perceives things in succession, and a third that perceives things in depth. These senses are sight, hearing, and touch. /Things alongside one another constitute a surface. Things in succession in their purest and simplest form constitute sounds. Things in depth are bodies or forms. Thus we have distinct senses for surfaces, sounds, and forms; and when it comes to beauty, we have three senses relating to

[7] Henceforth I refer to, but do not always follow for the sake of unity, Herder (2002a), Herder (2002b), and Herder (2006) as the English translations of Herder's writings.

[8] Although "force (Kraft)" is a problematic concept, Herder explained it just before the quotation above: "Painting operates in space and by means of an artful representation of space. Music and all energetic arts operate not just in but also through time, through an artful succession of tones. Would it not be possible to reduce the essence of poetry as well to such a primary concept, since it achieves its effect on the soul through arbitrary signs, through meaning of word? We shall call the means of this effect force" (194). For this concept in the context of aesthetics (focusing on Herder), see Menke (2012).

Table 3.2 Herder's trichotomy

Sight	Things alongside one another	Surface	Painting	Space
Hearing	Things in succession	Sound	Music	Time
Touch	Things in depth (ineinander)	Body	Sculpture	Force

three different genres of beauty that must be distinguished from one another just as we distinguish surfaces, sounds, and bodies. If there exist forms of art for which the proper domain is to be found in one of these species of beauty, then we know both their internal and external fields of application: on the one hand, surfaces, sounds, and bodies; on the other, sight, hearing, and touch. These limits or boundaries are imposed by Nature herself. They are not a matter of convention of agreement, and no decision can be made to alter them without Nature taking her revenge. Music which would paint, painting which would create sound, a sculptor who would employ color, a painter who would carve stone—all these are monstrosities irrespective of whatever effects they produce. All three arts are related to one another as surface, sound, and body, or as space, time, and force, the three great media of all-embracing Creation itself, through which they encompass and delimit everything there is. (Herder 1985–2000, IV 257)[9]

From the phrase "music which would paint, painting which would create sound" we can see Herder's ridicule on the color harpsichord. Indeed, he mentions it later in this work:

> Father Castel's ocular harpsichord has revealed to us sufficiently clearly what a fine art of colors for the sense of sight would be like and the sort of effects it would have. So many reasons have been put forward that are false or only partially true as to why this art did not succeed. The real reason, or at least the most natural, is that without the contribution of a more fundamental sense, the sense of sight affords us only a panel of light and color, and thus only the flattest and emptiest pleasure. A creature without hands who could see but could not feel forms or what is expressed through forms—in short, the head of a bird—might take pleasure in such an art, but no one else. (IV 279)

Herder did not consider sight "fundamental": For him, instead, touch, which "a bird" could not have, is worth being called "the most fundamental" (250).[10] This is because "[n]ature proceeds with each individual person in the same way as it does with the race as a whole: it moves from feeling to seeing, from sculpture to painting" (IV 302). Herder argued so by referring to (the consequence of) the so-called "Molyneux's problem" (245):[11] If a person born blind can feel the differences between shapes such as spheres and cubes, could (s)he similarly distinguish these objects by sight if given the ability to see? Since posed by William Molyneux (1656–1698) and introduced by John Locke (1632–1704) in *An Essay Concerning*

---

[9] Herder said that this book was "largely written in the years 1768–70" (244). Fourth Grove of *Critical Forests*, which was written during this time and published posthumously, contains therefore the corresponding contents and is, in a sense, a study work for *Sculpture*. For this argument see also II 307.

[10] See also II 325

[11] See also II 295f.. For the problem in general, see e.g. Morgan (1977).

Human Understanding (1690), this question had been heatedly discussed by philosophers such as Locke himself, Gottfried Wilhelm Leibniz (1646–1716), and George Berkeley (1685–1753). Then, Voltaire (1694–1778) reported in *Elements of Newton's Philosophy* (*Eléments de la philosophie de Newton*, 1738) the surgical recovery from blindness operated by William Cheselden (1688–1752) in 1729, according to which the patient could see, but did not know the shapes of things, and hence could not distinguish one thing from another.

Based on this, Herder discussed the relationship between the senses of touch and sight. How then is hearing related to them? To this question, *Treatise on the Origin of Language* (*Abhandlung über den Ursprung der Sprache*, 1772) replied: "Since the human being only receives the language of teaching nature through the sense of hearing, and without this cannot invent language, hearing in a certain way became the middle one of his senses, the actual door to the soul, and the bond connecting the other senses" (Herder 1985–2000, I 746). In a word, sight needs not only touch, "the most fundamental" sense, but also hearing to mediate and connect the senses. This argument is based on Herder's insight that "[w]e are a thinking sensorium commune, only touched from various sides" (743).

We, people in the modern age, tend to take the distinction among the five senses for granted, but Herder did not. According to him, the human being exercises their five senses together, based on that they are originally one and the same (which he called "feeling [Gefühl]" in a wider sense than "touch"), even in the developed and differentiated phase, where we, however, forget that "with much effort we learn to separate them in use" (744).[12]

So Herder criticized the color harpsichord also in this treatise: "Who always attentively gape ate a color clavichord without soon going blind?" (Herder 1985–2000, I 748). Now we can understand what his negative remarks on Castel's idea really mean. He criticized it not because it was an abortive attempt to force a connection between sight and hearing, two modally different senses, as his predecessors had done:[13] but instead because it lacks doubt or reflection on sight and hearing

[12] For this point, see Sugiyama (2009).

[13] Mendelssohn, whose close dialogue with Lessing produced *Laocoon*, said, for example: "It is indisputable that we are able to discriminate far more colors than sounds in the same time. For experience teaches that each individual color still sustains itself within the eye for a while if, at the same time, we have closed our eyes. Hence, in a melody of colors the impression which the foregoing colors have left behind them must be mixed with the present ones and produce an effect completely different from the one demanded. The nerves of the sense of hearing appear to retain the impression for a shorter time if the sound does not cause a very large vibration in the air. Indeed, although it may be hoped that, by habitualizing the nerves of the eye, one could even bring things to this point (since in the case of hearing itself it is very much a matter of habit), at least at the outset one would still have to let the colors follow upon one another more slowly and after lengthier intervals than do sounds, and only after long practice be mindful of a unification of the melody of the colors with the melody of sounds. / Father Castel first came up with the idea of putting a melody of colors into practice, and Krüger improved this invention considerably. But neither must have thought of this difficulty" (Mendelssohn 1929ff., I 117). It seems noteworthy here that Mendelssohn referred to "(a matter of) habit (Gewohnheit)" and "practice (Übung)". Therefore, he might have thought it possible in future—though impossible in his time—to accept (the mode of perception based on) the color organ without awkwardness, given long practice or habitualization of the nerves, whereas Herder thought it embryologically impossible.

being modally different and forces an unnatural mode of perception in reverse of the process of human development, from a "sensorium commune" to a being with five differentiated senses. In terms of synesthesia, he argued as a result that it cannot be artificially evoked (even by the color harpsichord), although, then, he could not know what contemporary neuropsychological research on synesthesia has since been clarified, i.e. that it occurs involuntarily.[14]

3.4 Conclusion

We can characterize Herder's aesthetics sketched above (from his remarks on the color harpsichord) as "sentio, ergo sum." Indeed, he wrote so twice in 1769: "there is nothing in the world of which I am immediately convinced by an inner feeling other than that I exist, I feel" (Fourth Grove of the *Critical Forests*: Herder 1985–2000, II 252) and "I feel myself! I am!" (manuscript titled "On the sense of touch [Zum Sinn des Gefühls]": IV 236). The Cartesian "cogito, ergo sum" became a starting point not only for modern philosophy through consideration of human spirits consisting in thinking but also for modern natural science by making the physical bodies, separated from spirit, calculable and observable. Now it is time to overcome (the view based on) this proposition. I think "sentio, ergo sum" should count as an alternative to "cogito, ergo sum." Many will endorse my proposal, such as the neuroscientist Antonio Damasio (1944–), who points out "Descartes' Error" (Damasio 1994) and one of whose book is translated into German with the title "Ich fühle, also bin ich" (Damasio 2000). For this view, Herder's aesthetics is the starting point.

References

Anonymous (1757) Explanation of the ocular harpsichord, upon shew to the public, London
Caduff C (2002) Fantom Farbenklavier. Das Farbe-Ton-Verhältnis im 18. Jahrhundert oder vom Einspruch gegen das clavecin oculaire und seinen ästhetischen Folgen. Zeitschrift für deutsche Philologie 121:481–509
Castel L-B (1723) Traité d'optique … par M. le Chevalier Newton; traduit par M. Coste. Journal de Trévoux 1723:1428–1450
Castel L-B (1725) Clavecin pour les yeux, avec l'art de peindre les sons et toutes sortes de pièce de musique. Mercure de France 1725:2552–2577
Castel L-B (1735) Nouvelles expériences d'optiques et et d'acoustique. Journal de Trévoux 1735:1444ff
Castel L-B (1740) L'optique des couleurs, fondée sur les simples observations et tournée surtout à la pratique de la peinture avec figures. Paris
Chouillet-Roche A-M (1976) Le clavecin oculaire du père Castel. Dix-huitième siècle 8:141–166
Cytowic RE (1993) The man who tasted shapes. MIT Press, Cambridge, MA
Damasio A (1994) Descartes' error: emotion, reason, and the human brain. Putnam, New York

[14] See e.g. Cytowic (1993, 73).

Damasio A (2000) Ich fühle, also bin ich. Die Entschlüsselung des Bewusstseins. Übers. von Hainer Kober. München: List

Descartes R 1996 ([1]1897–1913). Œuvres. 12 vols. Publiées par Charles Adam & Paul Tannery. Nouvelle éd. Vrin, Paris

Franssen M (1991) The ocular harpsichord of Louis-Bertrand Castel. The science and aesthetics of an eighteenth-century cause célèbre. Tractrix 3:15–77

Herder JG 1985–2000. Werke in zehn Bänden. Hg. von Günter Arnold et al. Deutscher Klassiker, Frankfurt am Main

Herder JG (2002a) Philosophical Writings (ed and trans, Forster M). Cambridge University Press, Cambridge

Herder JG (2002b) Sculpture: some observations on shape and form from Pygmalion's creative dream (ed and trans, Gaiger J). University of Chicago Press, Chicago

Herder JG (2006) Selected writings of aesthetics (ed and trans, Moore G). Princeton University Press, Princeton

Jewanski J (1999) Ist C = rot? Eine Kultur- und Wissenschaftsgeschichte zum Problem der wechselseitigen Beziehung zwischen Ton und Farbe von Aristoteles bis Goethe. Studio, Sinzig

Krüger JG (1740–1749) Naturlehre. 3 Bde., Halle

Krüger JG (1743) De novo musices, quo oculi delectantur, genere. Miscellanea Berolinensia ad incrementum scientiarum ex scriptis societati regiae scientiarum 7:345–357

Lessing GE (1968) ([1]1886–1924) Sämtliche Schriften. 23 Bde. Hg. von Karl Lachmann. ND. Berlin: de Gruyter.

Mendelssohn M (1929) Gesammelte Schriften. Jubiläumsausgabe. Akademie, Berlin

Menke C (2012) Force: fundamental concept of aesthetic anthropology (trans, Jackson G). Fordham Univeristy Press, New York

Morgan MJ (1977) Molyneux's question: vision, touch and the philosophy of perception. Cambridge University Press, Cambridge

Rousseau J-J (1959–1996) Œuvres complètes, 5 vols. Gallimard, Paris

Schier DS (1941) Louis Bertrand Castel, anti-Newtonian scientist. Torch Press, Cedar Rapids (IA)

Sugiyama T (2009) Herder's theory of common sense: the birth of the concept of synesthesia. Aesthetics 13:69–81. http://www.bigakukai.jp/aesthetics_online/aesthetics_13/text/text13_sugiyama.pdf. Accessed 30 June 2014

Telemann GP (1739) Beschreibung der Augen-Orgel oder des Augen-Clavicimbels, so der berühmte Mathematicus und Jesuit zu Paris, Herr Pater Castel, erfunden und ins Werk gerichtet hat. Hamburg

Chapter 4
A Brief Consideration About the Relationship Between Sound Art and Tactile Sense

Katsushi Nakagawa

Abstract In this chapter, in order to contribute to the aesthetics of tactile sense, I consider the relationship between sound art and tactile sense. First, I provide a historical perspective on sound art. I propose three periodical divides and six categories of sound art. Next, based on that historical perspective, I consider the function of tactile sense in this area. As my historical perspectives depend heavily upon the concept of the reduction of every art form into its own medium which was explored by Clement Greenberg, I conclude that there is not a large space for tactile sense in the history of sound art (tactile sense does not appear dominant there). After mentioning this position of tactile sense, I propose two ways of utilizing tactile sense in sound art works: as a metaphor and as an alternative means to listen to sound.

Keywords Sound art · Contemporary music · Contemporary art · Sound studies

4.1 Introduction

In this chapter, in order to contribute to the aesthetics of tactile sense, I consider the relationship between sound art and tactile sense. First, I provide a historical perspective on sound art. I propose three periodical divides and six categories of sound art. Next, based on that historical perspective, I consider the function of tactile sense in this area.

The roots of SA can be divided into three time periods: the latter half of the 19th century, in which it began to develop in two separate artistic fields (visual art and music); the mid-20th century, after WWII, when the center of sound art moved from Paris to New York; and the 1980s to the present day, during which time it constituted

K. Nakagawa (✉)
Institute of Urban Innovation, Yokohama National University, Yokohama, Japan
e-mail: katsushinakagawa@ynu.ac.jp

Y. Suzuki et al., *Computational Aesthetics*, SpringerBriefs in Applied Sciences and Technology, https://doi.org/10.1007/978-4-431-56844-5_4

an art form unto itself (Licht 2007). Thus, regarding the historical perspective of SA, I propose three periodical divides:

1. From 1850 to 1945
2. From 1945 to 1980
3. From 1980 to the present

I also propose six categories within those three periods, which are each divided into subcategories. The following is my historical perspective on SA, which I will explain in detail in the proceeding sections of this chapter.

1. From 1850 to 1945

   [1] New technologies and imagination

   (i) [1-1] Literature
   (ii) [1-2] New perspectives on technology: Reproduction as production
   (iii) [1-3] Experiments in sound film

   [2] Synesthesia between seeing and hearing
   [3] Origins

2. From 1945 to 1980

   [4] Context in music: SA as "new music"

   (i) [4-1] Emphasis on the visual element of music
   (ii) [4-2] "New" experimental music after John Cage
   (iii) [4-3] Attention to space/environment in music

   [5] Context in visual art: SA as "new visual art"

   (i) [5-1] Music by visual artists
   (ii) [5-2] Sound sculpture as the inclusion of sound in visual art
   (iii) [5-3] Attention to space/environment in visual art

3. From the 1980s to the present

   [6] SA themes: Sound or hearing

In 2013, the Museum of Modern Art (MoMA) in New York City had its "first major exhibition of sound art" as explained in McFadden 2013, "Soundings: A Contemporary Score," which presented work by 16 innovative contemporary artists working with sound. Though it was not altogether comprehensive, this exhibition was a testimony to the maturation of this field of artwork. As stated in the exhibition catalogue (McFadden 2013), "these artists approach sound from a variety of disciplinary angles—the visual arts, architecture, performance, computer programming, and music." That is, their works demonstrate the diversity of the themes of this genre through divergence in the artists' ages and approaches.

I will now provide a historical perspective on SA before the 1980s. After that, I will give brief consideration to the relationship between tactile sense and hearing sense in SA. I will propose two ways of utilizing tactile sense in SA and refer to some works of SA that relate to tactile sense.

4.2 A Historical Perspective on Sound Art

4.2.1 From 1850 to 1945

The crucial events of this period in the history of SA are twofold: the advent of phonography and the rise of artistic modernism.

Phonograpy was a form of technology that could record and reproduce sound, which later developed into media for distributing and listening to music: Leon Scott's phonautograph (1857), Thomas Edison's phonograph (1877), and Emile Berliner's gramophone (1887). This new technology stimulated the artistic imagination [1] and, though not by direct influence, inspired a body of literature [1-1] that included Lautreamont's *Les Chants de Malodor* (1868) (Kahn 1999: 5–6), Villiers de l'Isle-Adam's *L'Ève Future* (1886), and Raymond Roussel's *Locus Solus* (1914). These are the products of imaginations stimulated by the advent of photography, but not by the real experience of phonography, as the phonographs in those pieces of literature are nothing but a kind of fantasy. The imagination activated by phonography caused some thinkers to regard sound reproduction technology as a creative tool for artistic production [1-2], as represented by Rainer Maria Rilke in *Ur-Gersente* (*Primal Sound*) (1919), Lászlό Moholy-Nagy in *Produktion-Reprodukter* (*Production–Reproduction*) (1922), and so on. Moreover, the advent of new technology directly caused the advent of new forms of artistic creation.

The advent of sound film occurred in the 1920s (Katz 2004). Sound film was originally invented as a technology that added sound to silent films. Because sound film can be easily shaped—unlike the solid structure of phonographs or gramophones—one can edit its sound by reversing the track, changing the speed of play (and thereby the pitch), and transposing the position of the sound (mixing), as was done by Grigori Aleksandrov and Sergei M. Eisenstein in *Romance Sentimentale* (1930) and Walter Ruttmann in *Wochenende* (*Weekend*) (1930) [1-3]. These were innovations in sound editing that were attempted before the appearance of Pierre Schaeffer's post-war Concrete Music (Kahn 1999).

Artistic modernism—a progressive perspective on evaluating art that tries to acquire something unique, irreducible, and particular to it, such as the canvas of oil paintings in the mid-20th century—relates to [2] and [3]. The concept of the reduction of every art form into its own medium was explored by Clement Greenberg in his article "*Modernist Painting*" (first printed in 1961), and achieved its most prominent position in the 1950s with the abstract expressionist paintings of Jackson Pollock. From this perspective, [2] and [3] were the intermediaries between visual art and music. I call them "intermediaries" because they do not belong directly to the history of medium reduction, as they use more than two media in painting and music. We can find some examples of [2] in the abstract paintings of Kandinsky, who also expressed the synesthesia between seeing and hearing. Additionally, we should note the presence of experimental films (known as "absolute film," "abstract film," or "pure film") such as Hans Richter's *Rhythmus 21* (1921), Viking Eggeling's *Symphonie Diagonale* (1921) [Fig. 4.1], and Oskar Fischinger's *Studie Nr. 6* (1930) (Brougher and Mattis 2005). [3] also stems from this singular modernistic perspective.

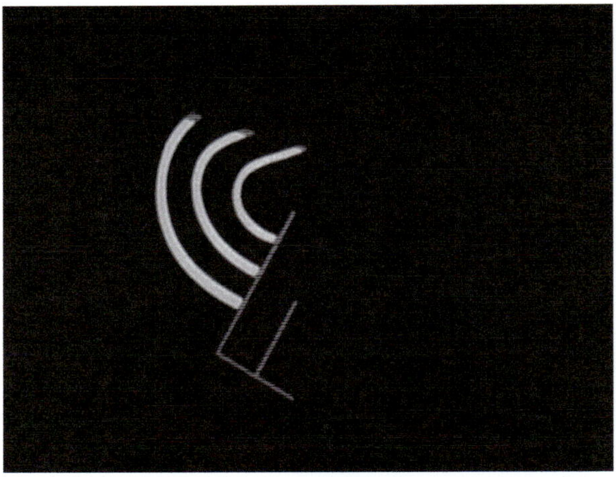

Fig. 4.1 Viking Eggeling, *Symphonie Diagonale* (1921) (Brougher and Mattis 2005: 102)

Fig. 4.2 Marcel Duchamp, *A Bruit Secret* (*Hidden Noise*) (1916) (Art or Sound 2014, unpaged)

[3] does not refer to the real origins of SA, but to what was set up as the "origins" of this history by posterity—a reversal from the way in which we talk about history nowadays. This is why there are several pathways into these "origins," such as Marcel Duchamp's conceptual works *Erratum Musical* (*Musical Erratum)* (1913) and *A Bruit Secret* (*Hidden Noise*) (1916) [Fig. 4.2]; the sound poetry of F. T. Marinetti, Hugo Ball, and Kurt Schwitters; the manifestations by Luigi Russolo (*The Art of Noise* and *Intonarumori* [Fig. 4.3]); and the early experimental music of

Fig. 4.3 Luigi Russolo's Intonarumori (Licht 2007: 97)

Erik Satie and John Cage (Cage 1961, Nyman 1974). I do not know if one can identify a well-organized relationship among these works, but since it was demanded that they serve as the "origins" of SA, they can be referenced together despite their unclear relationship.

4.2.2 From 1945 to 1980

The crucial point in this period of the history of SA is the maturation and diffusion of artistic modernism, both in visual art and avant-garde music. In other words, a "new" movement appeared apart from existing modernist art (abstract expressionist painting or minimal art, as well as experimental music influenced by John Cage), which was called "sound art [4 and 5]. The name was used to represent something new and unprecedented. The end of artistic modernism also meant the end of medium reduction and the rise of [6]. This outward orientation was found both in visual art and avant-garde music and developed into a major branch of SA. In short, these categories emerged as the result of the pursuit of a new art movement.

Regarding the situation surrounding music, we can extract three categories that relate to the pursuit of newness. First, the movement put emphasis on the visual element of music [4-1]. This led to the transformation of works of art that used only audio elements into works of art that also used visual elements. These included transformations from music to SA such as the picturization of musical staff notation as graphic notation (John Cage, Morton Feldman, Toru Takemitsu); visually interesting musical instruments (Harry Parch, Jo Jones); and visual elements of vinyl records

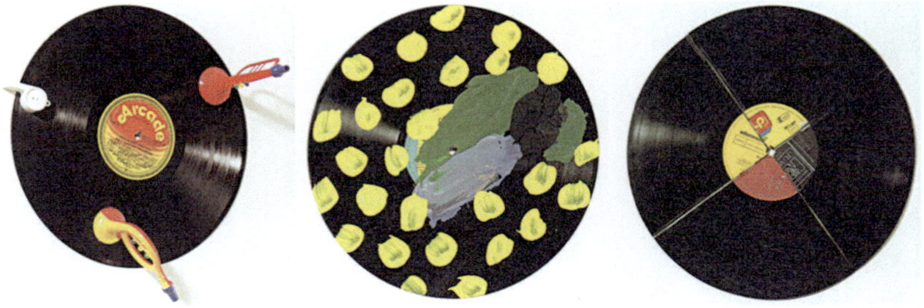

Fig. 4.4 Milan Knizak, *Broken Music* (1961-1979) (Block and Glasmeier 1989: 161)

(Milan Knizak, *Broken Music* (1961–1979) [Fig. 4.4], Christian Marclay, *Recycled Records* (1980–86), etc.). Second, experimental music modeled after that of John Cage spawned new ways of exploring music and demanded a new name that declared its own newness [4-2]. This included the sound installations of Bill Fontana, which he called "sound sculptures" (Fontana 1987) and the sound works of David Dunn after the 1970s (Dunn 1997). Third, the experimental and avant-garde music that came about after the latter half of the 1960s found new ways to progress towards acquiring outwardness; artists began to be interested in bringing the space or environment into musical works [4-3]. Works interested in the inclusion of space in music are represented by Karlheinz Stockhausen's *Gruppen* (1957) and Alvin Lucier's *Music on a Long Thin Wire* (1977), among others. Works interested in the inclusion of environment in music are represented by Erik Satie's *Musique d'Ameublement* (1920), John Cage's *4'33"* (1952), and others. This lineage was activated by R.M. Schafer's concept of the soundscape, developed in the 1970s. As a result, "sound installations" appeared, including Max Neuhaus's *Times Square* (1977–1992), Bill Fontana's *Satellite Soundbridge* (1987), Akio Suzuki's *Space in the Sun* (1988) [Fig. 4.5], and so on. Needless to say, this is entirely contrary to the concept of medium reduction, as this category tries to acquire something outside of itself and therefore has the capacity to include visual elements in originally audio works.

Regarding the situation surrounding visual music, we can also extract three categories. The first is [5-1] (Celant 1977, 1981). One reason why this category appeared, around 1960, is that sound recording technology became easily available after World War II, beginning with the magnetic tape recorder. Music by Jean Dubuffet and Joseph Beuys is representative of this category. Second, some visual art began to include aural elements, coming to be called "sound sculpture," though it differs from Bill Fontana's "sound sculpture" despite having the same name. [5-2] includes works such as Francois and Bernard Baschet's sounding objects (1952–) [Fig. 4.6], which were intended to be exhibited in public spaces; the sound sculptures of the designer Harry Bertoia (1957–), which the viewer can touch; the inclusion of sound into a museum exhibit in Jean Tinguely's *Homage to New York* (1960); and Robert Morris's *Box with the Sound of Its Own Making* (1961) [Fig. 4.7]. Third, avant-garde visual artists after the latter half of the 1960s also found new ways to progress toward acquiring outwardness. They began to be interested in bringing the

Fig. 4.5 Akio Suzuki, *Space in the Sun* (1988) (The Wire 2017)

Fig. 4.6 Francois and Bernard Baschet's sounding objects (1952-) (Licht 2007: 226)

Fig. 4.7 Robert Morris, *Box with the Sound of Its Own Making* (1961) (Art or Sound, unpaged)

space/environment into their works [5-3]. These works are referred to with the label "Earth Art" or "Land Art" and include Robert Smithson's *Spiral Jetty* (1970), Walter de Maria's *The Lightening Field, 1971–77, New Mexico* (1971–1977), and so on. As this category tries to acquire something outside of itself, it is in complete opposition to the concept of medium reduction and therefore has the capacity to include audio elements in originally visual works.

4.2.3 From 1980 to the Present

The crucial events in this period of the history of SA were the advent of the phrase "Sound Art" and the maturation of themes for this genre [6]. We can take the work of Yukio Fujimoto, such as *Ears with Chairs* (1990) [Fig. 4.8], as an example of this kind of SA. In this piece, the viewer settles down into a chair and adjusts two long tubes on each side of his/her ears, through which he/she listens to the drone of whatever room in which the chair is placed. The sound of footsteps, the hushed voices of other spectators in the museum, and the hum of air-conditioning equipment intensify, hollow out, magnify, and transform into a kind of drone. That is, the environmental sounds transform through this simple device. This provides an excellent opportunity for the viewer to find fresh attention in the environmental world around him/her. One can experience the world anew as something pregnant with significance. Each individual can decide whether or not this kind of music or visual art is impossible, unnecessary, or unfruitful, but it certainly seems that this type of art pulls its own weight as "good art." Through it, we can experience the world anew via the reconsideration of (environmental) sound and of our own hearing.

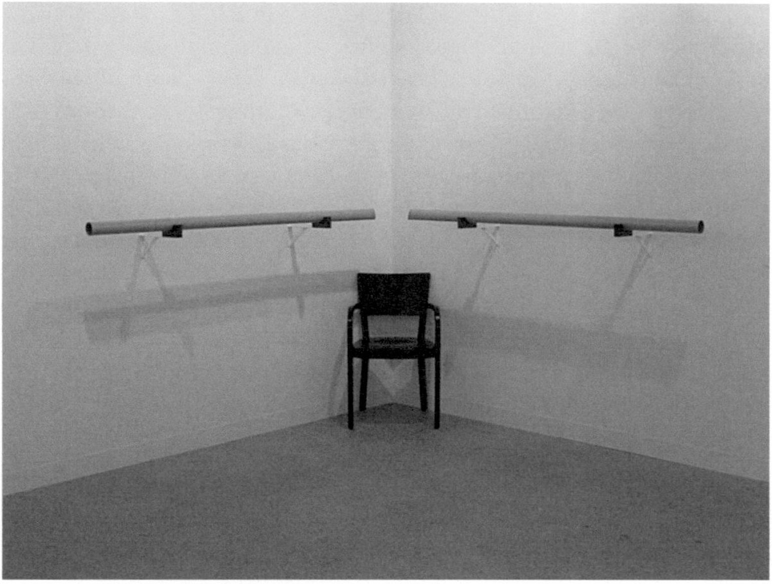

Fig. 4.8 Yukio Fujimoto, *Ears with Chairs* (1990) (ShugoArts 2017)

4.3 Tactile Sense in Sound Art

Now that we have examined the historical perspective of Sound Art, we can consider the function of tactile sense in this area—specifically, how we can contribute to the aesthetics of tactile sense?

The basic sense organs used in SA are seeing and hearing. We can look at the history of SA in terms of how the synesthesia between seeing and hearing has changed or developed. The 1983 exhibition catalogue that used the phrase "sound art" for the first time as its title stated the following:

> It may be that sound art adheres to curator Hellermann's perception that 'hearing is another form of seeing,' that sound has meaning only when its connection with an image is understood…The conjunction of sound and image insists on the engagement of the viewer, forcing participation in real space and concrete, responsive thought rather than illusionary space and thought. [Goddard 1983]

In the early stages of the history of SA, some themes in painting and music developed differently. Only later, around the 1980s, did they gather into one flow that constituted SA. This is the historical perspective on SA that I provide in this article; however, I cannot avoid concluding that there is not a large space for tactile sense in the history of SA, as tactile sense does not appear dominant in the history of SA.

In that case, how can we examine "tactile sense in SA"? First of all, we should understand that the tactile and hearing senses are adjoining, adjacent, and similar. Both senses can perceive vibration: touch perceives the vibration of objects, while hearing perceives the vibration of air. Moreover, according to the theories of embryology and evolution, hearing evolved and developed from touch when amphibians became mammals. I would propose that these facts explain why there is no trope of synesthesia between hearing and touch. They are not in a corresponding relationship, nor are they in a relationship that can be explained in terms of other perceptive senses. They are merely similar. While sight and hearing are not necessarily (or biologically) connected, touch and hearing are necessarily (and biologically) connected in nature. I believe that because of these facts, tactile sense can be used as a substitute or metaphor for hearing sense. Therefore, I want to propose that tactile sense is used in SA in two ways:

1. We use tactile sense as a metaphor: e.g., we "touch the sound."
2. We use tactile sense in order to listen to sound via touch, not via hearing, e.g., through bone conduction.

I will explain these proposals as follows.

Regarding the first point: we often use rhetorical phrases that originate from tactile sense for ordinary Western art music, such as a pianist's touch. These are metaphors, not the result of any kind of synesthesia. We also use phrases concerning tactile sense when we talk about the noise music of La Monte Young or other musicians, in which music physically moves the listener and where the listeners listen to music not only through hearing but also through touch. In this case, a phrase may be neither the result of synesthesia nor metaphors, but the bare description of the experience of the listener.

Regarding the second point: SA is not the effect of some kind of magic, but the utilization of the natural functions of the human body to use touch instead of hearing in order to listen to or perceive sound. Some works of SA use tactile sense as the composing element, such as Laurie Anderson's *Handphone Table* (1978) [Fig. 4.9]. Though Anderson is a musician, we can categorize this work as belonging to categories 5 or 6 because this work consists of a table and chair accompanied by sound. The visitor sits on the chair in front of the table, on top of which are two black circles where the visitor puts his/her elbows. Because the visitor must bend his/her back when putting his/her elbows on the circles, he/she must place his/her palms over each ear. The visitor then listens to the sound through his/her palms because the table conducts the sound through its legs. That is, sound flows through the table to the visitor's bones in a process of bone conduction. In this work, tactile sense is used as a substitute for hearing.

Another example is the conception of a "wave" by Takehisa Kosugi realized in his sound installation work such as *Catch Wave (mano-dharma electronics)* (1967)

Fig. 4.9 Laurie Anderson, *Handphone Table* (1978) (AHFM 2012: 230)

[Fig. 4.10]. Kosugi presumes that a "wave" has a ubiquitous existence as "the electric wave which is not visible to the eye and which an ear does not hear." This concept exists all over the world, inaudible and unseen. Kosugi regards sound as the externalized form of this "wave" and believes that this "wave" can also be actualized not only as sound, but also as wind, light, and so on. This conception can be interpreted as belonging to categories 4.1 or 4.2 because the conception was created through Kosugi's endeavor in order to overcome the influence of John Cage. In other words, this conception was created in the tradition of experimental music after John Cage. The "wave" can be interpreted as something like the environmental sounds in John Cage's work, which is also ubiquitous. The presumption of this ubiquitous existence can be interpreted as a strategy of listening to silence. If silence is ubiquitous, then it is simply inaudible sound, and one has only to make this sound audible; as a result, the desire for all sounds is satisfied (Nakagawa 2003, Kosugi 1968, 1991). In the case of Kosugi's "wave," we can listen to the vibration of the air—or, more precisely speaking, we can touch the vibration of objects. In short, we use tactile sense in order to perceive the "wave." Tactile sense is a substitute for hearing sense here as well. Here we can find the transformation of listening and music: we listen not only through hearing, but also through touch.

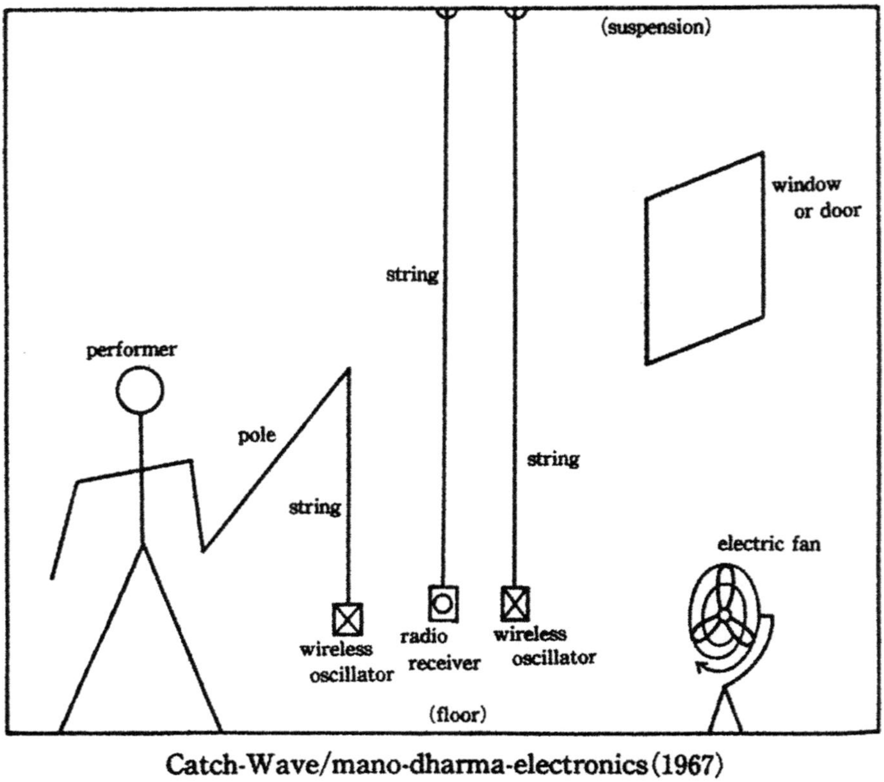

Fig. 4.10 Takehisa Kosugi, *Catch Wave (mano-dharma electronics)* (1967) (Kosugi 1991: 91)

4.4 Conclusion

This paper provides a historical perspective on SA and proposes two ways of utilizing tactile sense in works of SA. It does not provide a definitive principle for using tactile sense in SA, but it does provide examples for the consideration of the aesthetics of tactile sense. In SA, tactile sense is used as a substitute for hearing sense or as a kind of metaphor for hearing sense.

References

Beil R, Kraut P (eds) (2012) A house full of music: strategies in music and art [AHFM]. Exhibition Catalogue for the Institut Mathildenhc and Art 2012. Hatje Cantz Verlag, Darmstadt
Block U, Glasmeier M (1989) Broken music: artists' recordworks. DAAD and gelbe Musik, Berlin
Brougher K, Mattis O (2005) Visual music: synaesthesia in art and music since 1900. Hirshhorn Museum of Contemporary Art. Thames & Hudson, London
Cage J (1961) Silence. Wesleyan University Press, Middletown

Celant G (1977) The record as artwork: from futurism to conceptual art. Exhibition Catalogue. The Fort Worth Art Museum, Fort Worth

Celant G (1981) Artsound. In: Neuberger Museum (ed) Soundings: an international survey of sound in the plastic arts (trans, Weinberg CS). Neuberger Museum, New York, pp 4–6

Celant G et al (eds) (2014) Art or sound. Fondazione Prada, Milano

Dunn D (1997) Nature, sound art and the sacred. In: Rothenberg D, Ulvaeus M (eds) (2001) The book of music and nature: an anthology of sounds, words, thoughts. Wesleyan University Press, Middletown, pp 95–107

Fontana B (1987) The relocation of ambient sound. Leonardo 20(2):143–147

Goddard D (1983) Sound/art. In: Hellermann W, Goddard D (eds), Exhibition catalogue for "sound/art" at The Sculpture Center, May 1-30, and BACA/DCC Gallery, New York City, June 1–30, 1983

Greenberg C (1961) Modernist painting. In: Harrison C, Wood P (eds) Art in theory, 1900–1990: an anthology of changing ideas, vol 1993. Blackwell, Oxford, pp 754–760

Kahn D (1999) Noise, water, meat: a history of sound in the arts. MIT Press, Cambridge

Katz M (2004) Capturing sound: how technology changed music. University of California Press, Berkeley/Los Angeles

Kosugi T (1968) The look of sound: a slow interview with Takehisa Kosugi. Interview by Jud Yalkut. Art Magazine 42:14–15

Kosugi T (1991) Ongaku no picnic (musical picnic). Hakubasyobou, Tokyo (in Japanese)

Licht A (2007) Sound art: beyond music, between categories. Rizzoli, New York

McFadden S (ed) (2013) Soundings: a contemporary score. Exhibition Catalogue for the Museum of Modern Art, August 10–November 3, 2013. The Museum of Modern Art, New York

Nakagawa K (2003) On Kosugi Takehisa's Kosugi Takeh. In: Sasaki K, Otabe T (eds) The great book of aesthetics: proceedings of the 15th international congress of aesthetics, Japan, 2001 (CD-ROM)

Nyman M (1974) Experimental music: cage and beyond, 2nd edn. Cambridge University Press, New York 1999

ShugoArts (2017.) ShugoArts artists Yukio FUJIMOTO. http://shugoarts.com/old/artists/yukio-fujimoto/. Accessed 14 Apr 2018.

The Wire (2017) Akio Suzuki's space in the sun has been demolished. https://www.thewire.co.uk/news/49047/akio-suzuki-s-space-in-the-sun-has-been-demolished. Accessed 14 Apr 2018

Chapter 5
A Spectro-Gestural-Morphological Analysis of a Musical-Tactile Score

Eric Maestri

Abstract In the following chapter the author will compare a musical and a tactile score in the specific case study represented by Rieko Suzuki's tactile score transcription of Friedrich Chopin's posthume *Nocturne* n° 20 in *C sharp minor*. In the intention of the author this comparison will show its correlations between sound, movement and musical representation and will highlight their common features. To conduct this analysis, the movement indicated by the musical score and the ones prescribed by the tactile score will be compared using and expanding Denis Smalley's spectromorphological analysis towards a *spectro-gestural-morphological* analysis. For this reason, the author invite the reader to refer to Smalley's explanation of spectromorphological analysis (Smalley D, Organ Sound, 2:107–126, 1997). This research must be considered as a contribution related with the actual debate on sonification (Hermann T, Hunt A, Neuhoff JG, The sonification book, Logos Publishing House, Berlin, 2011) musical cognition (Launay J, Empiric Musicol, 10(1), 30–40, 2015, Schaefer RS, Empiric Musicol, 9(3–4), 161–176, 2014) and mediation (Leman M, Embodied music cognition and mediation technology, MIT Press, Cambridge, 2007) and will implicitly suggest a possible artistic application of tactile scores as basis of new compositional techniques.

Keywords Tactile scores · Musical scores · Morphology · Spectromorphology · Sonification · Haptic scores

5.1 Tactile Scores

Tactile scores were invented by Dr. Rieko Suzuki to describe and design tactile actions on the skin. These scores are used to notate actions oriented to massage to prescribe the movements to a masseur and to create a repertory of "pieces" that are performed on customers' visages. They are used to transcribe musical works in

E. Maestri (✉)
CICM (Centre de recherche en Informatique et Création Musicale),
University Paris 8, Saint-Denis, France
e-mail: eric.maestri@univ-paris8.fr

Y. Suzuki et al., *Computational Aesthetics*, SpringerBriefs in Applied Sciences and Technology, https://doi.org/10.1007/978-4-431-56844-5_5

massage-scores – in the case of tactile scores based on preexisting music – to compose music, to develop emotional engineering and haptic design (Suzuki and Suzuki 2014). Tactile scores are the result of a long process of development, based on trials and errors, that at defines the most important massage's parameters: i.e. trajectory of the massage on the visage, pressure and speed. During this period, Dr. Rieko Suzuki tried new massage's strategy with her customers developing more detailed scores: from hieroglyphic-like inscriptions to diastematic notations, she progressively reveals common features with musical scores expressing a certain proximity between the tactile and the musical prescription modality. Suzuki claimed:

> [...] counts and rhythm are important in tactile perception; a single circular stroke could not be distinguished from a mere rubbing, while more than double strokes would be recognized as massage. By giving rhythm on a tactile sense, we can create "impressions"; a rhythm of touching gives a "theme" on the impression provoked by tactile sense, where the theme is the expression through tactile sense such as small-large, fast-slow, line-curve and so on; in a sequence of massage strokes starts from small circles then moves to large circles and small circles again, à subject would feel small and large (Suzuki and Suzuki 2014: 32).

Tactile scores notate gestures characterized by specific movements of the hand on the visage; they are based on the profiles, speed and pressure changings made by the masseur. Thus, the succession of those tactile objects defines a temporal tactile objects that provoke complex sensations and feelings:

> We suppose that our image caused by tactile sense emerges from the temporal relationship; we always compare the tactile sense in the past and at the present. If we touch something hard, and then touch something harder, we regard the former as softer (Suzuki and Suzuki 2014: 32)

The development of tactile scores reveals some common features with musical scores. Indeed, as tactile scores show, the development of notations reveals the arousal of the awareness of a substantial temporal aspect of the syntax: as musical scores, tactile scores are based on objects that, for Suzuki, constitute a massage-alphabet. Tactile and musical scores are characterized by a similar graphical representation strategy, in both cases diastematic. Tactile scores provide an example of multi-modal musical-tactile representation and offer an important case study giving the possibility to compare actions related to sound perception. Indeed, tactile scores, recoding acoustic perception in tactile experience, provide an example based on the functionality of the gestural prescription through graphical inscription. As in music, in which notation prescribes the resulting sounds towards gestures, tactile scores, prescribe gestures to solicit tactile experience. These scores reveal "basic schemata of perception" (Godøy 2006) where gestures and body movements are a multi-modal crucial element. We will try to explain and analyze this interesting parallelism with Pierre Schaeffer's notion of "sonorous object" (Schaeffer 1966: 268). Thus, in the following of this chapter, the author will try to propose an enlargement of this notion into that of "sonorous-gestural-object" already proposed by Godøy, constituting the grounding notion of a perspective spectro-gestural-morphological analysis.

5.2 Gestural-Sonorous Objects and Spectromorphology

A brief analysis of Pierre Schaeffer's notion of sonorous object will show the utilization of gestural metaphors in the qualification of the sonorous objects. Sonorous objects are sound fragments extrapolated from the causal event and their sound agent. They could be defined by a random cut (*version*) or an intentional one based on the *stress-articulation* principle (*theme*) (Schaeffer 1966: 390–92). This notion allows to define the units that compose the auditory stream, offering a first historical attempt to define the auditory experience of a mediated perception. Sonorous objects provide

> the means for capturing and reflecting on the otherwise ephemeral or transitory nature of musical sound, i.e. allowed what we could call a mental recoding of sound to more stable images in our minds. (Godøy 2006: 149)

Actually, they provide a powerful analytical (and compositional) means, capable to recode the auditory experience in multi-modal images. Godøy stated:

> Schaeffer did make use of a number of gestural concepts and metaphors in qualifying sonorous objects. Going a bit deeper into this, I shall argue that Schaeffer's use of gestural concepts and metaphors can be related to the idea of so-called embodied cognition, meaning that virtually all domains of human perception and thinking, even seemingly abstract domains, are related to images of movement. For these reasons, I shall introduce the concept of gestural-sonorous objects in this paper, meaning an extension of Schaeffer's thoughts on exploring sonorous objects to also include the exploration of gestures associated with the various sonorous objects (Godøy 2006: 150).

Godoy hypothesizes that the strategy of segmentation applied for the analyses of sonorous objects could be applied in the analysis of gestural objects on the basis of a fundamental cognitive schemata:

> […] there is a continuous process of mentally tracing sound in music (and in musical imagery as well), i.e. mentally tracing the onsets, contours, textures, envelopes, etc., by hands, fingers, arms, or other effectors, when we listen to, or merely imagine, music. This means that from continuous listening and continuous sound-tracing, we actually recode musical sound into multimodal gestural-sonorous images based on biomechanical constraints (what we imagine our bodies can do), hence into images that also have visual (kinematic) and motor (effort, proprioceptive, etc.) components. Furthermore, this recoding is conceived of as a bidirectional process, i.e. that gestural images may engender sonorous images as well (Godøy 2006: 149–50).

Starting from this assumption, largely debated today, our analysis aims to dissect the relationship between the sound movements and the massage movements expressed in the tactile score. We will analyze the sound-tracing gestures (Godøy 2006: 154) in order to highlight the deeper cognitive structure expressed in the notation. This notion of gestural-sonorous objects will be used as starting point of an unorthodox application of the spectromophological approach proposed by Denis

Smalley, used as a methodological framework for the analysis of gestural-sonorous objects. Spectomorphology provides:

> [...] tools for describing and analysing the listening experience. The two parts of the term refer to the interaction between sound spectra (*spectro-*) and the ways they change and are shaped through time (*-morphology*). The spectro- cannot exist without the -morphology and *vice versa*: something has to be shaped, and a shape must have sonic content. Although spectral content and temporal shaping are indissolubly linked, we need conceptually to be able to separate them for discursive purposes – we cannot in the same breath describe what is shaped and the shapes themselves. The term may be rather jargonistic and it is perhaps an ungainly word, but I have not managed to invent an alternative which encapsulates the interactive components so accurately. Each component of the term belongs to other disciplines (visual, linguistic, biological, geological), which is fitting since musical experience radiates across disciplines. But the combination is unique: in music we often need words which are invented specially for defining sonic phenomena. A spectromorphological approach sets out spectral and morphological models and processes, and provides a framework for understanding structural relations and behaviors as experienced in the temporal flux of the music. (Smalley 1997: 107)

Spectromorhology is fundamentally linked to the referential activity of the listener, considered as inescapable for the any listening strategy. Indeed, this methodology defines the auditory stream perception as rooted in the listener's referential gestural experience, defined by Smalley as "source bonding". Smalley suggests that sound perception is linked with the agency's perception and with the gestural memory that links together the tactile and the acoustic experience. This approach expresses the notion of *sound imagery* (Launay 2015), linking spectromorphology to the so called embodied cognition. Indeed, in Smalley's words:

> One might think that in more 'abstract' instrumental music, source bondings do not exist, but they are there in force, revealed through gesture and other physical activity involved in sound-making. The bonding of instrumental activity to human gesture is somewhat ignored not only because it is invariably expected in music, but also possibly because much music study has tended to concentrate on music writing (notation), theory and analysis, which tend to distance a work from the gestural activity of its performance. Vocal presence, whether revealed through stylized singing or direct utterance, has direct human, physical, and therefore psychological links. In electroacoustic music the instrumental and vocal are included, but unlike traditional music where instrumental and vocal presence are assumed and known in advance, in acousmatic electroacoustic music they are both to be encountered (unexpectedly) and inferred. Electroacoustic music, then, subsumes instrumental and vocal experience: instrumental and vocal resources are a 'subset' of the wide-open sound world which is electroacoustic music's territory. (Smalley 1997: 111)

Smalley relates the perception of sound and its source and agent proposing an analytical method based on the constitutive distinction of *gesture* and *texture*, distinguished on the basis of listener's proprioceptive experience. This experience is provoked by the sonic traces of the human agency behind sound sources. Smalley claimed that:

> A human agent produces spectromorphologies via the motion of gesture, using the sense of touch or an implement to apply energy to a sounding body. A gesture is therefore an energy–motion trajectory which excites the sounding body, creating spectromorphological life. From the viewpoint of both agent and watching listener, *the musical gesture-process is tactile and visual as well as aural*. Moreover, it is proprioceptive: that is, it is concerned

> with the tension and relaxation of muscles, with effort and resistance. In this way sound-making is linked to more comprehensive sensorimotor and psychological experience. We should not think of the gesture process only in the one direction of cause–source–spectromorphology, but also in reverse – spectromorphology–source–cause. When we hear spectromorphologies we detect the humanity behind them by deducing gestural activity, referring back through gesture to proprioceptive and psychological experience in general. Everyone uses this spectromorphological referral process when listening to recordings of instrumental music. Not only do we listen to the music, but we also decode the human activity behind the spectromorphologies through which we automatically gain a wealth of psycho-physical information. [...] The basic gesture of traditional instrumental music produces the note. In tonal music notes form a consistent low-level unit, and are grouped into higher levelled gestural contours, into phraseological styles, which traditionally have been based on breath-groups. Singers and wind-players, after all, have to breathe. In electroacoustic music the scale of gestural impetus is also variable, from the smallest attack-morphology to the broad sweep of a much longer gesture, continuous in its motion and flexible in its pacing. The notion of gesture as a forming principle is concerned with propelling time forwards, with moving away from one goal towards the next goal in the structure – the energy of motion expressed through spectral and morphological change. Gestural music, then, is governed by a sense of forward motion, of linearity, of narrativity. The energy-motion trajectory of gesture is therefore not only the history of an individual event, but can also be an approach to the psychology of time. If gestures are weak, if they become too stretched out in time, or if they become too slowly evolving, we lose the human physicality. We seem to cross a blurred border between events on a human scale and events on a more worldly, environmental scale. At the same time there is a change of listening focus– the slower the directed, gestural impetus, the more the ear seeks to concentrate on inner details (insofar as they exist). A music which is primarily textural, then, concentrates on internal activity at the expense of forward impetus. (Smalley 1997: 113–14)

Sonorous and gestural objects emerge as expressions of a fundamental anthropological schema. The gestural perception of sound components allows the perceivers to understand the sound source's quality (the "*facture*" in Pierre Schaeffer's terminology) and the agency (the "*agent*"). This approach criticizes the notion of "reduced listening" that defines, in the terms proposed by Pierre Schaeffer, the very essence of the "sonorous object" as the correlative of reduced listening, conceived as a listening strategy based on the abstraction of the cause and sense from the sonorous objects. On the opposite, Denis Smalley proposed an embodied listening, rooted in listener's cognitive experience.

Our analysis will try to highlight some aspects of this complex relationship of sounds and gestures. It will concern the profile of the musical object and the global external morphology of both of them. We will use the definition of spectromorphological movement to compare the musical object trajectories in *registers* (the melodic profile) and in *agogic* (the dynamic profile); the tactile gestures will be analyzed via the movement of the *hand*, the movement of the hands on the *face* and the *pressure*. The movements of the melodic and the dynamic profile are considered as traditionally indicated in musical scores: ascending means from a lower to a higher register (melodic profile), or from weak to a strong one (dynamic profile). For the movements of the hand (on the face, of the hand, of hand's pressure) the ascension of the massage is meant from the lower to the upper part of the face, i.e. from the chin to the forehead, from the palm to the fingertips, from a weak to a strong pressure. Descending is defined as the opposite.

5.3 Analysis

In the next paragraph the author analyze the first phrase of the second section of Chopin's *Nocturne in C sharp minor* (opus *posthume*), completely transcoded in a tactile score (the bottom part of the score, Fig. 5.1). This musical phrase is played in the middle register and presents a very light and melodic atmosphere, in opposition with the darkest character of the first section. Before, we invite the reader to read Suzuki's article about tactile scores in which she traces tactile score's symbols and meaning (Suzuki and Suzuki 2014: 21–26). The first phrase of this second section is characterized by four sub-phrases that correspond to the bars. The next analysis will be based on the definition and comparison of the following gestural and sonorous movement typologies: *melodic profile* (the movement towards the registers); *dynamic profile* (movement of the dynamic); *hand movement* (the movement of the hand on its part); *face movement* (the movement of the hands on the face); *pressure* (hand's pressure).

In the first bar the musical unit is based on a movement from E_4 to D_5 and back to H_4, creating a *reciprocal-parabola* profile, characterized, dynamically, by a *crescendo* and *diminuendo (reciprocal-parabola).* This profile is defined by three types of motions. The first one is *unidirectionnal-ascending* (from E_4 to D_5); the second one is *cyclic/centric* (around the $C\#_5$) and slightly *descending* to the final note. About the hand, the tactile score indicates a movement from the palm going progressively to the fingertip (*unidirectional-ascending*); the movement on the face is of type *unidirectional-plane*, fixed on the chick; the indicated pressure is from half-pressure to full-pressure (*unidirectional-descending*), indicating an augmenting stronger pressure.

Second bar is characterized by a *unidirectionnal-(descending)* (from E_5 to H_4) and *cyclic/centric* motion – compared to the first bar – that finishes on the H_4, that is the central pitch of the first two measures. This musical profile is characterized by a small ornamentation around the final note. The dynamic is stable (the same of the previous bar), characterized in the third and fourth beat by a *diminuendo*

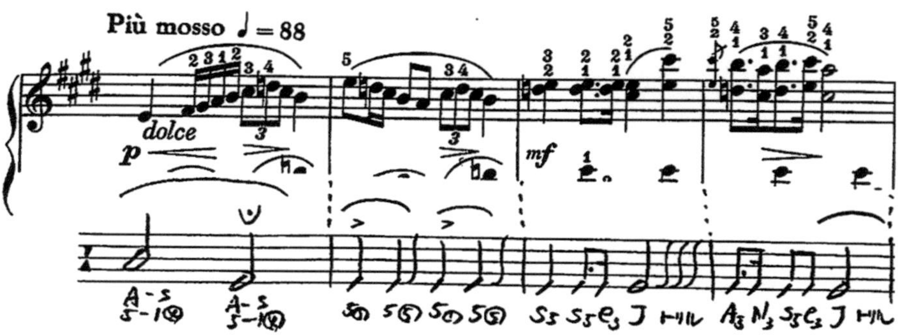

Fig. 5.1 Chopin's Nocturne opus posthume in C Sharp Minor recoded in a tactile notation (system in the bottom)

(*unidirectionnal-descending*). The movement of the hand is plane (*centric-pericentrality*) while the movement on the face is fixed on the forehead (*cyclic/centric*). The pressure is characterized by two opposite movements: in the second and fourth beat the prescribed movement is palm-fingertip (fast lightening of the pressure) (*reciprocal-parabola* like: from the palm to fingertip and vice-versa); the first and third are characterized by an "accent" movement with the palm, defined by a stronger pressure on the chick (*centric/cyclic-plane*: stable on the palm).

On bar three the musical score indicates a periodic iteration in the higher register around D_5 and E_5, defining a *unidirectional-plane* movement, ascending in the second part of the measure with a bi-chord $C\#_5$-E_5, $C\#_6$-E_5 (*unidirectional-ascending* movement); the dynamic profile is *unidirectional-plane*. The tactile score follows the music in a homo-rhythmical manner, designing a *plane unidirectional* movement, in the first part of the bar (first two beats). In the first two bars the movement on the face is fixed on the forehead (*centric*), moving towards the eyes on the semi-quaver (beat two) (*unidirectional-descending*) and on the downbeat of the third beat on the chin (*unidirectional-descending*). Thus, the movement on the face defines a profile from the forehead to the chin passing through the eyes (it corresponds to a global *reciprocal peri-centric* movement from the chick to the chin). The movement on the surface of the *hand* is defined by a *centric-plane* movement, fixed on the palm. From the point of view of the pressure the movement is *plane* (first and second beat) and *iterative-unidirectional* (continuous movement from full to light pressure at each quaver of the last two beats).

The last bar of the phrase, is characterized by a *cyclic/centric-centrifugal* motion, from the point of view of the melodic profile, and by a *unidirectional-descending* dynamic profile. The tactile score indicates a movement on the face from the chicks (*cyclic-rotation*) to the chin through the neck, the forehead and the eyes (*unidirectional-motion*). This movement indicates a composite directionality that encompasses the whole face. The movement on the surface of the hand is *reciprocal-parabola*, between the palm and the proximal phalanges. Differently from the previous bars, the movement on the surface of the face and of the parts of the hand is parallel. The chick is massaged with the palm, the neck with the proximal phalange, the forehead with the palm and the chin with the proximal phalange. The pressure movement is *cyclic-centric-rotation*.

As schematically synthetized in Table 5.1, the tactile and the musical movements define a gestural-sonorous polyphony characterized by two clearly divided layers. The musical movements are defined by a fundamental inherent isomorphism, e.g. the first bar is characterized by a melodic and dynamic profile *reciprocal-parabola*. On the contrary, the tactile movement of the first bar is heteromorphic. Indeed, it is characterized by a *unidirectional-ascending* movement of the hand, a *cyclic/centric* face movement and a *unidirectional-descending* pressure. In the second bar the unidirectional movement of the musical movement is associated with *cyclic/centric* tactile one; the third bar is characterized by a *unidirectional-plane* sonorous movement and a multifaceted tactile movement, defined by a *centric-plane* hand movement, *reciprocal-pericentric* face movement and *unidirectional-plane* pressure. This polyphonic movement is characterized in the fourth bar by a *cyclic/centric*

Table 5.1 Comparative scheme of the musical and the tactile score

	Musical score		Tactile score		
Movements/ bars	Melodic profile	Dynamic profile	Hand movement	Face movement	Pressure movement
1	reciprocal-parabola	reciprocal-parabola	unidirectional-ascending	cyclic/centric	unidirectional-descending
2	unidirectional-descending	unidirectional-descending	cyclic/centric	cyclic/centric	reciprocal-parabola /centric/ cyclic-plane
3	unidirectional-plane	unidirectional-plane	centric-plane	reciprocal peri-centric	unidirectional-plane
4	cyclic/ centric-centrifugal	unidirectional-descending	reciprocal-parabola	cyclic-rotation/ unidirectional-motion	cyclic-centric-rotation

sonorous movement in the melodic profile and a unidirectional descending dynamic profile; the tactile object is defined by a *reciprocal-parabola* hand movement; a *cyclic/rotation* unidirectional face movement and a cyclic/centric pressure.

This analysis shows the very complex relationship between the gestures used to produce tactile and acoustic experience. The sonorous objects and the gestural objects have a strong internal coherence. E.g. the sonorous objects are isomorphic the melodic and dynamic profile. Indeed, they are characterized by a similar movement typology: *reciprocal*, *unidirectional* etc. On the opposite, gestural objects are defined by a kind of internal polyphony. E.g. hand *unidirectional* movement, face movement *cyclic* and *unidirectional* pressure; *cyclic* hand and face movement and *reciprocal-parabola* pressure. Massage's movements are characterized by a detailed differentiation caused by the extremely fine capability of our tactile sense to discriminate slightly different informations detecting micro-changing in the massage texture. For that reason, it is difficult to highlight constants among these dimensions. However, the proposed methodology allows to define the correlations between sonorous and gestural objects and points, in the future, a class of gestural-sonorous objects.

5.4 Conclusion and Future Developments

This study allows to define the gesture's notation as a common element of the musical and the tactile scores. Indeed, it is clear that the notation emerges as a graphical representation of gestures, revealed as the central aspect of a cognitive schemata capable to recode, through gestural proprioceptive metaphors, the sensitive experience. Indeed, tactile experience is transposed in musical experience and vice-versa using gestural metaphors that transcode the informations coming from touching in multiple modals outputs. Gesture prescription appears as a crucial aspect of the representation of the temporal experience. Future developments are characterized by the improvement of the proposed analytical methodology and by the analysis of different case studies to have a larger number of examples capable to support the research in musicology, cognitive science, informatics and art. Finally, the author, that is a music composer, think that, on this basis, it will be possible to realize new pieces starting from tactile scores. This artistic research could be characterized by the definition of spectromorphological correlations between tactile and acoustic experience using Denis Smalley's theoretical framework to define the sound typologies of new musical pieces, especially in the domain of electroacoustic music.

Reference

Godøy R-I (2006) Gestural-sonorous objects: embodied extensions of Schaeffer's conceptual apparatus. Organ Sound 2(2):149–156

Hermann T, Hunt A, Neuhoff JG (2011) The sonification book. Logos Publishing House, Berlin

Launay J (2015) Musical sounds, motor resonance, and detectable agency. Empiric Musicol 10(1):30–40

Leman M (2007) Embodied music cognition and mediation technology. MIT Press, Cambridge

Schaefer RS (2014) Mental representations in musical processing and their role in action-perception loops. Empiric Musicol 9(3–4):161–176

Schaeffer P (1966) Traité des objets musicaux. Essai Interdisciplinaires. Editions du Seuil, Paris

Smalley D (1997) Spectromorphology: explaining sound-shapes. Organ Sound 2:107–126

Suzuki Y, Suzuki R (2014) Tactile score. A knowledge media for tactile sense. Springer, Tokyo

Chapter 6
Think of Everything with Back Calculation: Computational Aesthetics of Hair Design

Shinya Tsuchiya

Abstract Hair design, which has been developed by S. Tsuchiya is typical example of Computational Aesthetics, where all the details are algorithmically designed.

Keywords Hair design · Principle of hair design

6.1 What Is Hair Design?

Human can only move within the predetermined constraints with skeletons, joints, and muscles, and the constraint varies from person to person such as the length of limbs, the size of face, and the length of neck.

The slender proportion with long limbs and small head like a model is not absolute aesthetic shape. Hair design indirectly shapes the individual constraint to get in tune perfectly with the hairstyle. For example, we can't create the same beauty of a model with another person even if we design the hair to fit the person into the hairstyle of the model as a template.

The shape and size of face and head is different from person to person. The fundamental skeleton can't be changed. The hair grows covering the skull.

Regardless of age, sex, thick, or thin, any body it's self is already beautiful. The hair grows covering the skull, but the shape of the skull it's self has aesthetic appearance, so the basic of all the hair designs is the shape of the skull. Even if someone has long hair, the long hair is not beautiful which doesn't make others feel the person's skeleton.

S. Tsuchiya (✉)
ZA/ZA Co., Ltd, Tokyo, Japan
e-mail: shinchan0105.zaza@gmail.com

Y. Suzuki et al., *Computational Aesthetics*, SpringerBriefs in Applied Sciences and Technology, https://doi.org/10.1007/978-4-431-56844-5_6

6.2 Calculation Only by "Subtraction"

Hair design can only cut hair in principle, so in this regard, it is similar to carving. For example, to fluff hair out in any part, we can't add hair there, so reduce hair on another part and relatively make it seem to fluff. If you would like to emphasize something, you should relatively subtract other things.

If you would like to hide something, you can relatively cancel out by increasing it. For example, when a part of hair is thin and the other part is thick, the thin area stands out. In this case, reduce the hair of other part and increase thin area then you can make the thin hair less noticeable. If you have a part to put in the shade [in most cases that is face], design the hairstyle to shift one's vision to other parts. Therefore, we change the place to be gazed at by creating the highlighted area daringly and change the balance.

Cut can only do subtraction. The advanced hair design only uses subtraction for increasing volume of some parts like addition or spatially combining the parts by cutting hair to increase or decrease volume like multiplication (Fig. 6.1).

6.3 How to Make It "Halts"?

The way of hair design is similar to carving in forming objects by only cutting in principle. However, the completely different point is that hair grows and the shape changes depending on the angle of face and wind. Even though the object moves or the hair grows to some extent, the beauty of the hairstyle must be maintained.

Both carving and hair design form objects by cutting however the desired technique of carving is like hitting a ball at rest as golf, while on the other hand the desired technique of hair design is like hitting a moving ball as baseball or tennis.

Hair design can't choose materials as carving. It shapes hundred thousand hairs by changing the length or volume on each zone, and the combination of shaping has infinite possibilities.

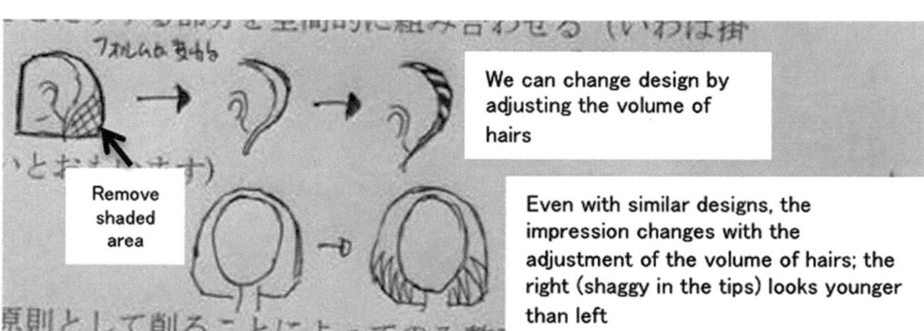

Fig. 6.1 (Upper) Design by changing a feeling of massiveness of form. (Lower) Even though the same form, it looks younger by adjusting the hair volume because the eye level is lifted

Fig. 6.2 Left and right figures are the same hairs, hairs move in any direction; hundred thousand hairs by changing shapes on each zone are moving and the combination of shaping has infinite possibilities

Hairs move in any direction, so they can't be formed in the same way as cutting stones or clay. And also, they are already dimensionally formed by skull, so the constraint is large (Fig. 6.2). Hair design is classified roughly into the following two categories.

1. The design taking no thought for the shape of head,
2. The design taking into account the shape of head.

Here we mainly describe about 2. Hairs fall down to the direction of gravitational force along skull. They move by turning the face or wind, and the face rarely stays in a certain direction in daily life. Therefore, the hairstyle needs to be designed in advance assuming that hairs move affected by gravity and swing back and forth and around at about 45 degrees (e.g. Fig. 6.2, left).

We don't break hairs into chunks in hair design. We make hairs 3-dimensional as if folding each hair which falls down to the direction of gravitational force.

When cut hair longish around the top of head, it covers the upper part of head, at this time cut hair around neckline as supporting upper side hair from below, the design has volume at the bottom.

However, we have to consider customers' preferences such as the part of fringe, braids, or wisps with hairdressing. In the case of carving, appreciators don't change the sculpture after the completion however in the case of hair design, it is needed to design for maintaining beauty at a certain level even though the customer changes the style. This point is also widely different from carving.

The combination of the existence of hair is the design in hair design. Even though hair grows with time, the balance of the combination doesn't change. However, the place where hair grows down depends on gravity and the shape of individual head, so the hairstyle changes with time keeping the balance of the combination.

6.4 "Divide and Conquer Algorithm" in Hair Design

Divide and Conquer Algorithm is popular as one of classical algorithms. This algorithm solves a complex problem by dividing it into some small problems and combining them again.

The algorithm of my hair design is as follows.

1. Divide hairs into sections (in my mind).
2. Make sure the facial contour and decide the overall form such as roundness or vertically long shape.

3. Decide the design feature in detail such as fringe, neckline, sideburns, and the movement of the top. Image how the impression of detailed designs as a whole will be and how the customer wishes to be. I do this process before cutting while I image more concrete way for implementation.
4. Decide the design concept. The design concept means the overall form as the standing posture especially looking from side and back.

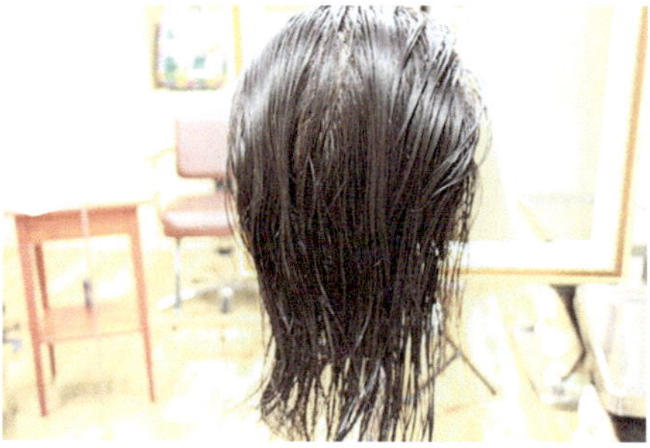

5. Convert each section of No.1 into design. Therefore, divide the head into three parts. One third is frontal of the head and two-thirds is the center and back of the head.

6. Design and balance each section.

7. Consider a part where the slope of hair is maintained virtually constant along the head to be a section. Cut the tip of hair or change hair volume lifting the hair in normal direction against scalp. In concrete terms, adjust hair density (Fig. 6.1).

8. Cut the lifting hair in each section counting backward the form when the hair falls down along the shape of the head. (Fig. 6.3)

Divide areas of the head into some blocks. The dividing way depends on aimed designs or hair conditions of customers. I change the cut way and hair volume in each section for creating design. The upper area in each section is influential, and the form of the design will change a lot if the line of the section is misaligned in one centimeter even though the cut line is the same (Fig. 6.4).

6.4.1 Balancing "Hidden Vectors"

As described above, I design hairstyle by sectioning hair and adjusting individual sections however I can't design with only the shape of hair.

I consider all things including the shape of face, hair tips, and the flow of hair as vectors (Fig. 6.6).

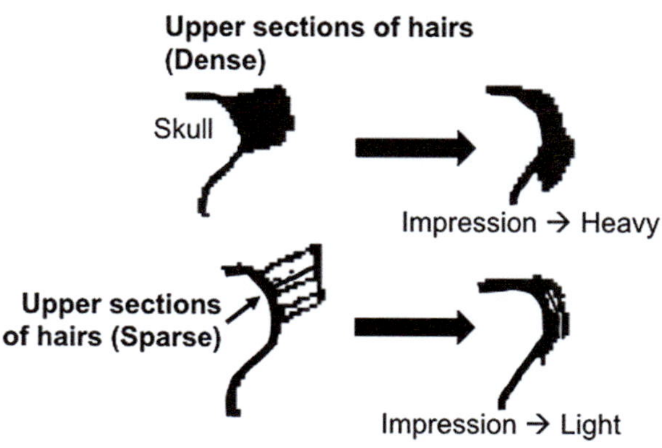

Fig. 6.3 Adjustment of hair amount in each section

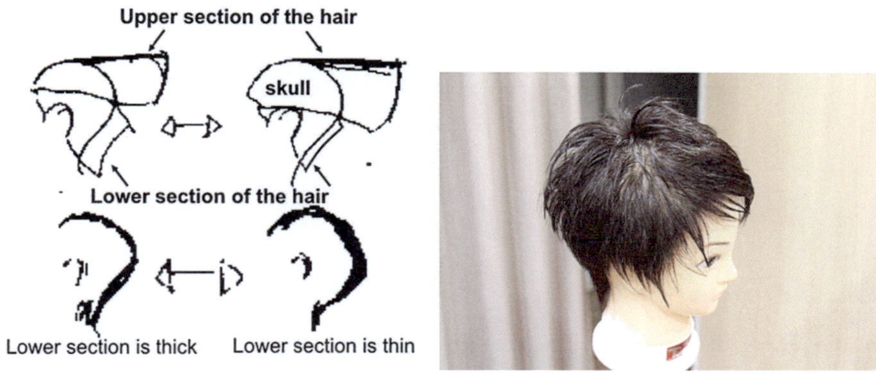

Fig. 6.4 Design determination in each section; the shapes of vertically upper sections are affected by the shapes of vertically right below sections; we will only see the outermost shape and it is formed by the accumulation of layers of vertically sections; likewise building an architecture, we have to design from the innermost sections and accumulate each layer

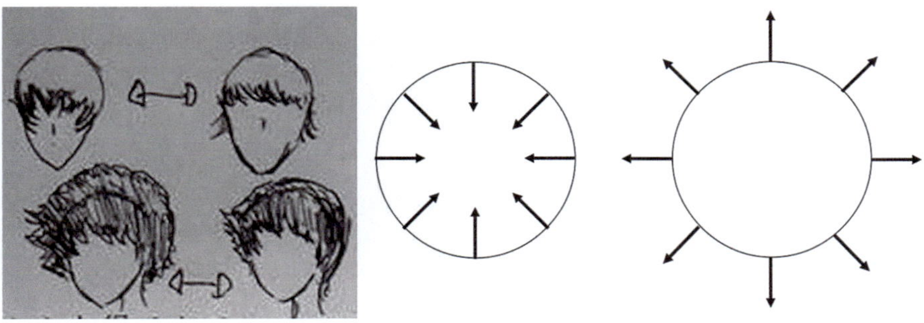

Fig. 6.5 Gestalt of hair design. All faces are the same in size and shape while have different impressions depending on the hairstyle. Impressions of shapes convey by the vectors; although the size of cycles left and right are the same, right one looks large

The biggest vectors are the shapes of face and head. If the face is vertically or horizontally long, I assume that those vectors are from the center of face to longitudinal and horizontal direction respectively. They are fixed vectors and I can't change the vector itself. On the other hand, the directions of hair tips and flow of hair are vectors that I can handle.

Cognitive psychology (Gestalt psychology) (Soegaard 2012) shows that our perception recognizes not only absolute physical quantity (Fig. 6.5). Therefore, hair design can change the impression derived by irremovable vectors such as the shape of face with bundles of the vectors created by the shape of hair.

If the face is big, the impression of the face size can be looked smaller by directing hair tips to the center of the face or increasing hair volume and enhancing contrast between the face and the volume of the hair (Fig. 6.6).

Fig. 6.6 Flow of hair as vectors; directions of tips of hairs correspond to the directions of vectors and they create impressions; the size of face is the same but impressions (e.g. size of face) look different

6.5 Thinking Reversely

As I've discussed, I design the objective form by cutting like carving. Therefore, it isn't possible to shape the objective form directly, so divide it into some parts. It is required to think back ward from combining divided forms in a multi layered for creating the objective form. This approach is similar to origami paper or carbon board that is designed as a development figure backward from shaping it into three dimensions.

Additionally, in the case of hair design, the shape of hair changes with temporal and spatial variation by the growth of hair and the movement of face, so it is required to design a development figure in consideration of the design in four dimensions including time passage.

Epilogue: From the Stand Point of Computational Aesthetics

Yasuhiro Suzuki

There are two big points in general design in connection with methodology of hair design by Mr. Tsuchiya. One is product design whose concepts or functions are pre-established. In this case, the direction and object are clear, so a designer just designs toward them.

The other is the design by combining some elements, so that means it isn't a grand design. For example, sun-dried bricks are subtly different in color and shape one by one. A wall made by a craft person heaping up sun-dried bricks has unique texture however this texture can't be designed in advance.

In the case of hair design, there are two design ways as above however ordinary people don't clearly understand the difference. For example, hair moves by walking or shaking one's head, so does it mean that these movements may create unique texture? We think of dreadlocked or afro hair as fixed hairstyle whose texture of the form is created by assembling small elements.

Mr. Tsuchiya's model of carving hairs is controlled by evolving emergent computation; in his algorithm, carving hairs is constructing an architecture; firstly, the outer shape is designed and the structural calculation is performed and then he obtains construction drawings and starts cutting hair. However, hair design is fatally different from architecture, each of elements is moving and time developing. His works (hair designs) are instauration arts; they are exhibited on physically moving galleries, his customers; they are real instauration arts that participate to complete his art work every day by arranging their own hair style and exhibit the art work in public.

From Computational Aesthetics point of view, Algorithm of hair design, which is controlled emergent evolutional computing, is one of the most advanced algorithm and many useful algorithms are hidden there.

Reference

Soegaard, M (2012) Gestalt principles of form perception. Interaction-design.org. Retrieved 6 Apr 2012

MIX
Papier aus verantwortungsvollen Quellen
Paper from responsible sources
FSC® C105338

If you have any concerns about our products,
you can contact us on
ProductSafety@springernature.com

In case Publisher is established outside the EU,
the EU authorized representative is:
Springer Nature Customer Service Center GmbH
Europaplatz 3, 69115 Heidelberg, Germany

Printed by Libri Plureos GmbH
in Hamburg, Germany